LONG-TERM HAZARDS FROM ENVIRONMENTAL CHEMICALS

A ROYAL SOCIETY DISCUSSION

ORGANIZED BY SIR RICHARD DOLL, F.R.S.,
AND A. E. M. McLEAN
FOR THE ROYAL SOCIETY'S STUDY GROUP
ON LONG-TERM TOXIC EFFECTS

HELD ON 13 AND 14 DECEMBER 1977

LONDON
THE ROYAL SOCIETY
1979

Printed in Great Britain for the Royal Society
at the
University Press, Cambridge

ISBN 0 85403 110 3

First published in *Proceedings of the Royal Society of London*,
series B, volume 205 (no. 1158), pages 1–197

Published by the Royal Society
6 Carlton House Terrace, London SW1Y 5AG

CONTENTS

Contents

Proc. R. Soc. Lond. B. **205**, 3–4 (1979)
Printed in Great Britain

Preface

BY SIR RICHARD DOLL, F.R.S.

*Department of the Regius Professor of Medicine,
Radcliffe Infirmary, Oxford, U.K.*

All who are concerned with the effects of man's activities on the environment and the reaction of these effects on man must feel some disquiet at the number and amount of new chemicals that have been introduced, and are continuing to be introduced, into the environment. This disquiet led internationally to the establishment of the United Nations Environmental Programme and, in this country, to the establishment of the Royal Commission on Environmental Pollution and a Committee of Inquiry to review the provisions for the safety and health of people at work – the former under the chairmanship of Lord (then Sir Eric) Ashby and the latter under the chairmanship of Lord Robens. By 1972 both of these bodies had concluded that there was an urgent need to develop a system that would ensure that some assessment was made of the impact on man and his environment of all new chemicals, and the waste matter produced in their manufacture, before they were used on any large scale.

Since then, developments have been rapid. In April 1973 the Royal Society held a discussion meeting to consider the idea, suggested by the Royal Commission, for establishing a data bank of chemical substances and their environmental effects. The idea was approved and criteria were proposed that would determine which substances should be included (Cook & Warner 1974). One year later the Health and Safety at Work Act was passed and screening procedures, comparable to, but less stringent than, those applied to chemicals intended for use in food, drugs, and agriculture, were extended to substances for use in the place of work. Exactly what these procedures will be has not yet been decided, but proposals, which will form the basis of draft regulations, were submitted by the Health and Safety Executive for comment by interested parties in May 1977.

This Discussion Meeting owes its origin to discussions between representatives of the Royal Commission and the Royal Society in June 1974. A suggestion was then made that the Society should establish two Study Groups to consider:

(i) atmospheric chemistry, including the long-range dispersal and absorption of pollutants in the atmosphere and their effects on the environment, and

(ii) the long-term biological effects of toxic substances in low concentrations.

The Study Groups were duly appointed and, after reviewing the many problems within their remits, arranged two meetings. The first, on 'Pathways of Pollutants in the Atmosphere', was held in November 1977. The second is reported here.

The subject of this meeting is intended to include not only chemicals that have been manufactured by man but also substances, like heavy metals, that have been redistributed in the atmosphere and on the Earth's surface as a result of man's activities. We have excluded the effect of chemicals to which man is exposed as a result of smoking, drinking, and eating – even though deleterious effects may be grossly increased by industrial processes which alter the natural balance of foodstuffs or facilitate overconsumption – as these have been fully discussed individually on other occasions. Such chemicals are, however, included if they are added to preserve, to colour, or to give taste to food, rather than to improve nutrition.

The object of the meeting is to seek ways in which screening tests can be used to protect man and his environment from real hazards without at the same time being so restrictive that technological innovation is inhibited and social benefits fail to be obtained. This means that we must either know enough about the mechanism of disease to be able to predict from theoretical considerations not only what sort of effect, but also what incidence of effect, will be produced by exposure of a given number of people in a given way to a given amount of a particular chemical, or we must be able to validate the screening tests that we use by correlating the results with human experience. For it is only on this basis that government can make a sensible estimate of the balance between benefits and costs. We shall, therefore, in the course of our discussion, review the present state of health of the country and the extent to which it has been damaged by man-made chemicals inside and outside the workplace. It must, however, be remembered that small doses of some chemicals may take many years to exert their maximum effect and that current experience is not necessarily a reliable guide to the future.

The papers, it will be noted, deal almost entirely with the direct effect of chemicals on man and pay little attention to their effect on other organisms in the biosphere. This is not because these other effects are unimportant, but because the whole biosphere would have been too big a subject to tackle in a two-day meeting. We have, however, included two papers on hazards to wildlife, both because they illustrate clearly some of the problems of prediction and because they remind us of the complexity of the food chain on which man is dependent.

REFERENCE (Doll)

Cook, J. & Warner, F. 1974 Comments and conclusions. *Proc. R. Soc. Lond.* B **185**, 221–224.

Proc. R. Soc. Lond. B. **205**, 5–15 (1979)
Printed in Great Britain

Sources and extent of pollution

By Sir Frederick Warner, F.R.S.

Cremer & Warner, 140 *Buckingham Palace Road,*
London SW1W 9SQ, U.K.

Before the sources and extent of pollution can be identified a definition of pollutants has to be agreed. The degree of disruption of natural cycles in the global ecosystem in terms of residence times and assimilation capacities must be assessed as a prerequisite of any system of control.

The sources of man-made and naturally occurring chemicals that fall into this definition can be categorized and these are presented for reference. Specific examples of these categories are discussed in detail, e.g. sulphur dioxide, polychlorinated biphenyls and radioactive waste.

Their distribution and dilution in the environment are governed by fluid mixing mechanisms. These can be modelled to allow prediction of effects at specific points taking into account disappearance by decay, chemical reaction and deposition. Reappearance through pathways which involve accumulation and remobilization can only be predicted when a complete scientific understanding of the mechanism has been established.

I had hoped in preparing a background introduction to this symposium to review, among other things, the work of Professor Sugden's group on the atmosphere. Recent months have instead been completely filled with the task of hearing evidence about reprocessing nuclear fuel at Windscale, so that it was not possible to attend Professor Sugden's meeting and to take a deep look at my topic.

On the whole, it does not greatly matter because any attempt to cover the subject runs into trouble for lack of information and statistics. There is more on some matters than others and a study of some which have shown secular change may shed light on those that have come into prominence more recently, many of which indeed are to be reviewed in the papers that follow. It is only 7 years since the Royal Commission on Environmental Pollution began its work under Lord Ashby and the reports have been almost exclusively concerned with the pollution of rivers and estuaries, of the atmosphere and, in the 6th report, with problems associated with radioactivity.

Water pollution

The study of water pollution was concerned almost entirely with the effects of deoxygenation due in the main to sewage. The 3rd Report dealt with the polluting effects of chemical industry also in these terms and table 1 shows the comparisons made by the Royal Commission to show improvements in effluents in terms of biochemical oxygen demand.

[5]

This illustrates the first point to be made in any study of sources and extents. The figures available are given in terms of concentrations or indices, but the important thing is mass, and for this the gaps in information are great. The

TABLE 1. EFFECT OF NEW PLANT ON THE POLLUTION INDEX

(Pollution index as biochemical oxygen demand in kilograms per tonne of product.)

process	old plant (1960)	new plant (1970)
methanol	13	1.4
terephthalic acid	13	1.5
ethylene	1.3	0.22
ammonia	3.8	0.33

TABLE 2. CONURBATIONS, POPULATIONS AND RIVER FLOWS

river	city	population (millions)	daily flow/(m³/s) max.	min.	mean
Seine	Paris	3.0	2100	18	273
Elbe	Hamburg	2.0	3620	145	700
Danube	Vienna	1.7	9600	504	1916
Thames	London	10.0	360	1	62
St Lawrence	Great Lake area	30.0	9000	4600	7000
Potomac	Washington, D.C.	3.6	?	20.3	370

second point is the weakness of control systems which try to place limits on concentrations, because they are easily measured, where it is more important to know the mass discharged and the pattern of dilution and disappearance in the medium receiving it.

The differences in dilution which various river systems can afford is shown in table 2.

It does not show the Rhine whose flow has fallen at times to 1000 m³/s at Dusseldorf but is controlled by the Delta scheme in Holland so that the maximum flow through Rotterdam does not exceed 6000 m³/s. The Thames by contrast has to provide the main drinking water of London of 25 m³/s on average by abstraction above Teddington, with a requirement that 9 m³/s must pass over the weir to keep the estuary purged. In dry summers such as 1975 and 1976 this flow could not be sustained and fell at times to nothing. All of the drinking water which passes through London's people and industry runs back into the Thames estuary as sewage treated in some way, and the effectiveness of that treatment over the years up to 1970 is shown by the oxygen sag-curves in figure 1.

These curves show how oxygen-saturated fresh water flowing over Teddington Weir is first depleted and then recovers in the tidal estuary. The curves are average and demonstrate trends. Over short periods, trends have reversed, for example in November 1970 when there was a strike of sewage workers. This particular deterioration inspired the gloomy subsection in Jay Forrester's world model on pollution and the time for its decay. This form of Thames pollution in general disappears by the action of oxygen imported with tides, by reaeration from wind action and by physical transport to the sea when flow increases. The ranges of flow in normal conditions give a residence time in the estuary from 1 to 50 days. At long residence times even cellulose from paper mills is oxidized.

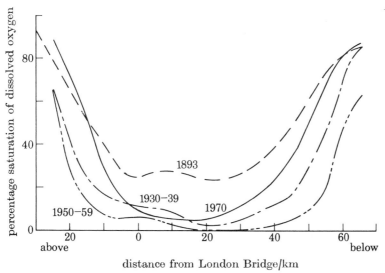

FIGURE 1. Thames estuary: dissolved oxygen levels.

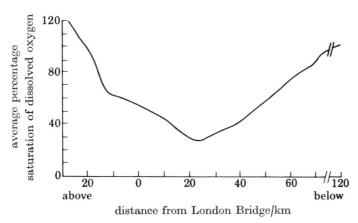

FIGURE 2. Thames estuary: dissolved oxygen levels, July to October 1976.
Flow rate: 1.2 m³/s.

Thames pollution was finally overcome by completion of sewage treatment works in 1975. After two dry summers, the estuary retained a minimum 30 % of dissolved oxygen in the third quarter of 1976 (figure 2) and after winter rain had recovered by the spring of 1977 to over 60 % (figure 3). The other evidence of recovery in terms of returning fish is a well known story.

AIR POLLUTION

The air pollution studies of the Royal Commission centred on sulphur dioxide
and smoke as their principal indicators. The changes which have taken place in
London between 1955 and 1974 are shown in figure 4. These again are average

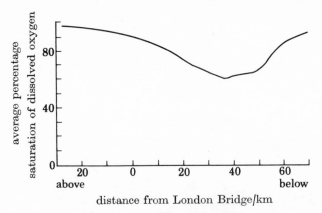

FIGURE 3. Thames estuary: dissolved oxygen levels, January to March 1977.
Flow rate: 196.1 m³/s.

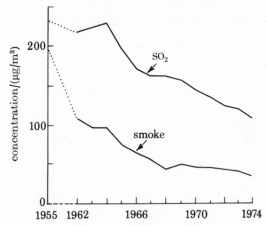

FIGURE 4. Annual averages of smoke and sulphur dioxide concentrations near ground level
in the City of London. From *Annual report of the Scientific Adviser to the G.L.C.*, 1974.

figures and it is necessary to look at extreme conditions for more effective illus-
trations. The impetus to air pollution studies was given by the inversion episode
of December 1952, which led eventually to the Clean Air Act. It is shown in
table 3 against comparable later inversions.

There was no episode in 1976 and so far none in 1977 but the 1975 episode
recorded SO₂ about 2000 µg/m³ on an hourly average, but it passed unnoticed
or at least without public comment. Health records for this period were not
available because of non-cooperation by medical staff in the hospitals.

SO$_2$ has a short residence time in the atmosphere, about 2 days, and concern has shifted recently to the resulting sulphate aerosols and acid mists which are washed out by precipitation. This topic has been discussed at Professor Sugden's study group and was dealt with at the Discussion Meeting in November 1977.

TABLE 3. MAJOR AIR POLLUTION EPISODES IN LONDON

(Concentrations are maximum mean daily concentrations in Central London.)

	concentration/(μg/m³)		estimated extra deaths in Greater London
month and year	smoke (B.S. method)	SO$_2$	
Dec. 1952	6000	3500	4000
Dec. 1962	3000	3500	750
Dec. 1972	200	1200	Nil
Dec. 1973	230	620	Nil
Dec. 1974	180	500	Nil

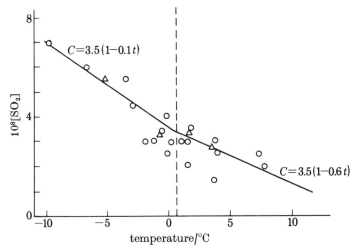

FIGURE 5. Sulphur dioxide concentrations in Stockholm. The broken line represents the climatological winter mean temperature for Stockholm.

It is documented in the O.E.C.D. study which shows that half the countries studied received the major part of their deposition from SO$_2$ of foreign origin, particularly in areas of heavy precipitation, such as southern Scandinavia and Switzerland. In general, the concentrations and resulting total depositions are a maximum in the major emission areas and decline with increasing distance from them. It gives no encouragement to proposals for setting standards for concentration in air, but focuses again on the problems of the mass involved in relation to circulation patterns.

These problems are illustrated by the local effects shown in the next three figures. Figure 5 shows how SO$_2$ varies with temperature in Stockholm. Figure 6 shows normalized SO_2 concentrations in Swedish cities plotted against the number of inhabitants. Figure 7 expresses a similar relation where the plot of SO_2 is against the notional city diameter.

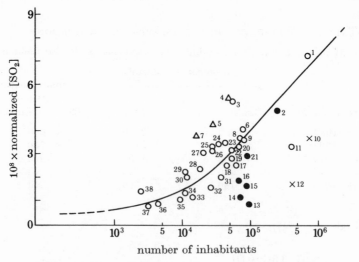

FIGURE 6. Normalized SO_2 concentrations in Swedish cities.

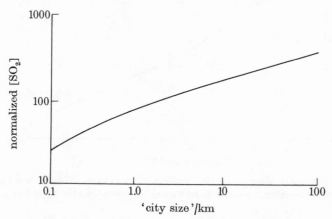

FIGURE 7. The relation between normalized SO_2 concentrations and notional city diameter.

The discussion is complicated by the variation in concentration with the time taken for sampling. The criteria used in different countries are illustrated in figure 8. The 5th Report of the Royal Commission suggested that there could be merit in setting guidelines or bands for air pollution control. Figure 9 shows an attempt to put figures on such bands for sulphur dioxide.

The air over the U.K. appears now to be clean enough to allow ozone to form. In June–July 1976 high concentrations (7200 parts/10^9 and over 8 h periods of

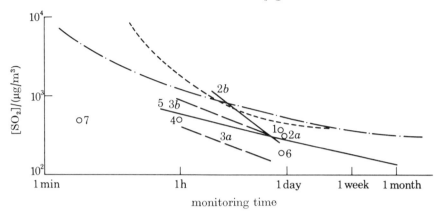

FIGURE 8. Comparison of national ambient air quality standards for SO_2. —·—, Human response curve; ---, vegetation curve. Standards: 1, Dutch; 2*a*, U.S. (primary); 2*b*, U.S. (secondary); 3*a*, Canadian (desirable); 3*b*, Canadian (acceptable); 4, Japanese; 5, Swedish; 6, W.H.O.; 7, U.K. 1 part $SO_2/10^6$ = 2860 µg/m³.

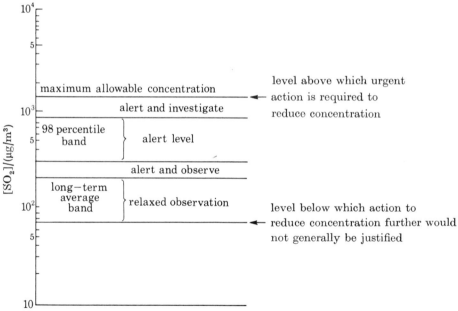

FIGURE 9. Possible air quality guidelines for sulphur dioxide, on the basis of 1 h sampling time.

1000 parts/10^9) were recorded at five sites (three in London). These were associated with air masses passing over southern Poland 50 h previously, across central Germany and up the Thames estuary.

Global estimates of air pollutants have been made in a number of studies by, among others, O.E.C.D., Stanford Research Institute and the Scope publications. Individual pollutants in the environment are covered by the publications of the Central Unit on Environmental Pollution.

Low concentration pollutants

Much of current concern over pollution has moved away from the classic areas taken by the Royal Commission to discharges from the chemical industry. The principal target is the chloralkali industry which has been at the source of many new products. It is an industry which fosters, and is at the mercy of, rapid change. Because brine must yield on electrolysis equimolar amounts of caustic soda and chlorine, a balance has to be found if market shares are unequal. This has meant developing new outlets for chlorine by associating the maximum amount in some organic compound such as chlorethanes, vinyl chloride or benzene hexachloride. A large industry has arisen for making solvents, pesticides, herbicides and plastics. Some of these have been associated with occupational diseases and make any chlorinated hydrocarbon suspect, particularly if it also has one or two benzene rings. Others have been found to damage fish and birds as a result of bioaccumulation and persistence in the environment. Their manufacture and use has been stopped or they have been kept out of the environment, for example by restricting polychlorinated biphenyls to dielectric and heat transfer fluids. Their persistence can be demonstrated by analytical techniques which can measure very small amounts. The detection of halogenated compounds at concentrations around 10^{-12} is only paralleled by the sensitivity with which radioactivity can be measured.

Pesticides and insecticides are applied by spray which is eventually transferred to soil, from which it evaporates and adds to the air-borne undeposited fraction or is adsorbed on solid particles. The larger molecular mass chlorinated compounds are relatively insoluble in water (10^{-7}) and are not transferred to percolating water. Some are degraded, if slowly, under aerobic and anaerobic conditions as well as photochemically in the presence of electron donors. The levels found in rain and rivers are around 10^{-12} and decreasing.

Small molecular mass compounds, on the other hand, are much more soluble in water (150–8000 parts/10^6) and more volatile. They are also, though not so strongly, absorbed on soil and are absent from underground waters. In rivers and reservoirs, the levels are orders of magnitude higher (10^{-8}). It has been suggested that some are of natural origin.

A recent example of an accidental escape of material which caused concern was at Seveso, where trichlorophenol was made as a herbicide. The batch reaction for manufacture can get out of control and 2,3,7,8-tetrachlorodibenzoparadioxine produced (TCDD or dioxin). At Seveso the reactor contents were discharged by bursting of a rupture disk to the outside air in a few minutes. The resulting cloud of trichlorophenol, ethylene glycol, caustic soda and TCDD was dispersed downwind and part deposited on the ground and other surfaces. Modelling studies of the dispersion and deposition to be expected in the prevailing atmospheric conditions showed that some 12–20 % of the reactor contents would be deposited in the zones characterized as highly contaminated, the remaining 80–90 % being dispersed over much wider areas at low levels. Figure 10 shows the isopleths calculated from modelling the dispersion of 2 kg of TCDD and their relation to the zones marked A and B as contaminated. The calculated levels match well those observed. Again their detection at such levels has come from the ability to combine chromatography with mass spectrometry in analysing mixtures in picogram (i.e. 10^{-12} g) quantities.

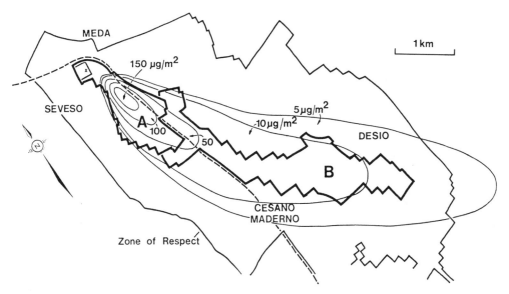

FIGURE 10. Seveso: contaminated zones. Calculated deposition of TCDD (2|kg release; case A).

There are not many examples where sources can be located, masses of discharge specified and then dispersed into an environment which has been sufficiently described. The preparation of the Thames Estuary model took a great deal of experimental work and few similar models exist. It proved also to be simplified because the sea water is completely mixed with the fresh water on each tide. Many estuaries are layered and the description is complicated.

Even with these physical advantages, the model is idealized and of utility only over long periods. The whole sewerage system which provides the inputs is also unique in sorting out many pollutants which are of individual concern. The mercury which enters the system is attached to suspended matter and its dispersion is not simply by dilution. In all systems, the role of dusts and sediments is becoming important and their transport in air or water complicates the simple models. This was noted earlier in respect of sulphur dioxide appearing as particulate sulphates and acid. The large molecular mass hydrogenated hydrocarbons which volatilize from soil reappear on smoke particles over cities.

RADIOACTIVITY

One closing example is in the area of radioactivity and comes from evidence recently given at the Windscale Inquiry into transuranics associated with sediments. Since 1974 these have been sampled by the Fisheries Radiobiological Laboratory at 26 stations in the Irish Sea and 18 close to Windscale and analysed for Pu, Am and Cm. Surface sediments in the Ravenglass estuary have also been analysed to check routes back to man. Concentrations of ^{241}Am between 40 and 250 pCi/g have been recorded between 1971 and 1975. Among possible pathways investigated was mussels near Windscale where the concentration factor from filtered water was $1-2 \times 10^3$ for Pu and $5-10 \times 10^3$ for Am. There was a correlation with discharges of 1–5 pCi/g respectively for 1 Ci/day discharged.

Air sampling took place during the Inquiry at the Ravenglass estuary because of suggestions that silt resuspended in air could be significant. Rough tests over long periods with tacky shades did not indicate this and high volume sampling showed that ^{239}Pu and ^{240}Pu averaged 2.2×10^{-4} pCi/m^3 and ^{241}Am 1.8×10^{-4} pCi/m^3 over the period.

A similar check on possible accumulation in soil dressed with seaweed in the Isle of Man showed amounts at the limit of detection, such as 0.35, 0.05, 7.2 and 10 pCi/kg for ^{239}Pu and ^{240}Pu in potato flesh, peel, seaweed and soil and 0.08, 0.08, 0.04 and undetectable for Am. These are recent figures for a discharge and are given for their current interest but without comment.

I conclude this introduction by reemphasizing the lack of quantitative information on sources, in spite of the large numbers of figures which are now available because of sensitive analytical techniques.

Discussion

SIR DOUGLAS BLACK (*Royal College of Physicians*, 11 *St Andrews Place, London, NW1 4LE, U.K.*). I suggest that action on environmental pollution raises considerable problems relating to the agencies which might be involved. The ultimate responsibility clearly lies with 'Government', in the interests of the general public. But 'Government' is a blanket term for politicians, and civil servants, both administrative and professional. Of these categories, politicians are vulnerable to the activities of pressure groups, which tend to concentrate on single issues, not necessarily the most important. Administrative civil servants have a strong concern for the public interest, taken broadly; but in scientific matters they depend on professional advice, though the ultimate decision often lies *de facto* with them, even if *de jure* it is for the politician. The professional scientist in a government department has a taxing role – not perhaps too dissimilar from that of the University teacher, in that he has to explain complex matters to highly intelligent men, largely without scientific preconceptions, but also liable to move around at short intervals. Not all of the problems of arriving at sound statutory provisions have been solved.

Manufacturing industry also has a clear responsibility, which is often exercized in a most responsible fashion, but cannot in the nature of things be totally free from bias. The popular media have influence, but like politicians they are liable to capture by pressure groups.

For these reasons, there is a strong case for independent expert scrutiny, under the supervision of bodies such as the Royal Society. Given sufficient vigilance, they are untrammelled by secrecy, by political consideration, or by commercial interests, and might thus be uniquely well placed to consider the public welfare.

Proc. R. Soc. Lond. B. **205**, 17–30 (1979)
Printed in Great Britain

Pollution of the sea and its effects

By H. A. Cole†

Formerly Director of Fishery Research, Ministry of Agriculture,
Fisheries and Food, Lowestoft, Suffolk, U.K.

Marine pollution has been studied under the following groupings of effects:
harm to living resources, hazards to human health, reduction of amenities
and interference with other users of the sea. This paper is concerned
mainly with the first two categories and their interrelation.

Apart from certain seabirds affected by oil, the major stocks of marine
animals show no evidence of reduction by pollution. Pollution effects are
generally insignificant in relation to other factors governing reproductive
success, survival, growth and population size. Even in the North Sea,
which has received a greatly increased pollution load during the last
three decades, both total production of fish and catch per unit of effort
(a measure of abundance) of cod, haddock and plaice increased during
the 20 years 1950–69. Very recent decreases have been due to overexploitation but, except in certain estuaries and immediate coastal
waters, direct damage by pollution to marine populations and ecosystems is not evident. Pollution effects can, however, be detected by
chemical analysis.

The paper examines human health risks arising in the marine environment, particularly from contaminated seafood, expecially in relation to
sewage pollution, metals such as mercury, cadmium and lead, synthetic
organic substances and oil.

Introduction

Pollution of the marine environment has been studied internationally under the
following groupings of effects: harm to living resources and marine life, hazards
to human health, reduction of amenities, and interference with other legitimate
users of the sea (Gesamp 1973). We are concerned here mainly with the first two
categories of effects.

It is very difficult to establish effects of pollution on natural populations of
marine animals and plants in the open sea. These populations are naturally subject
to wide variations in abundance, both from year to year, according to survival of
young, and in response to longer cycles of change, e.g. in climate. Moreover,
populations of fish, shellfish, whales, seals and other resources show marked
responses to exploitation. 'Fishing' pressure is high throughout the world and
the effects of overexploitation are evident in many species. It is, however, agreed
that, apart from certain seabirds affected by oil and possibly by accumulated fat
soluble substances, the major stocks of marine animals show no evidence of

---

† Present address: Forde House, Moor Lane, Hardington Mandeville, Yeovil BA22 9NW,
U.K.

reduction by pollution. This does not imply that pollution has no effect but that such effects are not significant in relation to other environmental factors governing reproductive success, survival, growth and population size. For a general account see Ruivo (1972).

In the North Sea, which has undoubtedly received a heavily increased pollution load during the last three decades, including a full range of synthetic organic chemicals, both the total production of fish, and the catch per unit of effort (a measure of abundance) of staple food species, such as cod, haddock and plaice, increased very markedly during the 20 years 1950–69 (Cole & Holden 1973). These increases have been shown to be due mainly to the occurrence of a number of very rich year classes. Even in the Baltic Sea, where increasing deoxygenation has occurred in the bottom waters due in part to pollution, the fisheries as a whole increased in productivity during the same 20-year period.

Very recent declines in the production of some species of fish, e.g. herring, from the North Sea are due to overexploitation. Specifically, they result from the deployment of too many powerful and efficient vessels which together have generated substantially more fishing effort than is required to harvest the maximum sustainable yield from the stocks. Pollution played no significant part in their decline (fishing for herring in the North Sea was banned in 1977) but it is conceivable that the presence of pollutants could have some delaying effect on recovery when excess fishing pressure is removed. Pollution must be regarded as one of a number of environmental factors, but certainly not the most important in an open sea situation, which may affect the survival of young fish. It must be remembered that although in human beings we are accustomed to regarding the individual as important, in fish there is generally a vast overproduction of eggs and larvae of which more than 99.9 % do not need to survive to maintain the population level.

Locally, in estuaries and coastal waters, fishery resources have been reduced by pollution, principally by the discharge of untreated sewage and other wastes with a high biological oxygen demand. These have the effect of reducing the area of suitable habitat for particular species. Static molluscan shellfish, such as oysters, clams and mussels, and fish and crustacea tied to an estuarine habitat during part of their life cycle (e.g. smelts and shrimps), have been particularly affected. In the U.K., fisheries production has declined markedly over the last 60 years in a number of estuaries as the result of industrial and housing development and increasing pollution. Good examples are provided by the Thames, Humber, Forth and Mersey but many smaller estuaries have also been affected. Nevertheless, inshore fisheries as a whole in the U.K. are in a thriving condition, and certain species with specialized habitat requirements have shown a surprising resilience. For example, since about 1970 there has been a spectacular re-development of fishing for oysters (*Ostrea edulis*) in the Solent (Key 1972, 1977). This area, probably because of the presence of a large oil refinery and much shipping, has been regarded by some as threatened by pollution but, in fact, must be classed as rather lightly affected (see Helliwell & Bossanyi 1975). It is especially interesting that the presence of warmed water from the power stations at Marchwood and Fawley (labelled 'thermal pollution' in the United States) seems likely to have been one of the factors creating environmental conditions conducive to the survival and setting of oyster larvae during the critical initial phases of the recovery of this fishery.

PERSISTENT POLLUTANTS

Although pollution effects cannot be detected in production levels of sea fisheries (except locally in respect of some estuarine fish and shellfish), evidence of pollution is easily found by the analytical chemist in enhanced levels of metals and organic substances resistant to biodegradation. The levels of some of these

TABLE 1. MEAN CONCENTRATIONS OF HEAVY METALS IN FISH AND
SHELLFISH FROM THE NORTH SEA

(Concentrations are in milligrams per kilogram fresh tissue;
ocean water concentrations are in micrograms per litre.)

	Hg	Zn	Cu	Cd	Pb†
cod	0.03–0.48	2.4–7.0	0.26–1.1	0.02–0.5	0.1–3.0
plaice	0.02–0.26	3.7–8.0	0.22–1.8	0.02–0.18	0.1–2.61
herring	0.02–0.24	3.0–17.0	0.6–3.6	0.02–0.7	0.2–5.1
mussel	0.06–0.68	18.0–33.0	1.7–13.0	0.17–0.5	0.8–7.2
shrimp	0.07–0.23	6.3–40.0	7.2–31.0	0.1–1.0	0.1–6.8
ocean water	0.15	2	2	0.11	0.03

(Data from I.C.E.S. (1974).)
† Some upper lead values are reported as uncertain.

TABLE 2. CONCENTRATION OF POLLUTANTS BY SHELLFISH (SEASONAL MAXIMA)

(*a*) Metal levels in shellfish compared with surrounding seawater

oysters	Cd	Cu	Pb	Zn
Ostrea edulis	103.8	597	10	3090 µg/g
Crassostrea gigas	136.2	853	10	6060 µg/g
sea water (filtered)	5.7	10.7	3.3	183.2 µg/l
concentration factor				
O. edulis	1.82×10^4	5.58×10^4	3.03×10^4	1.69×10^4
C. gigas	2.39×10^4	7.97×10^4	3.03×10^4	3.31×10^4

(Adapted from Boyden (1975).)

(*b*) DDT and PCB concentrations in mussels (nanograms per gram)

date	location	DDT	PCB
1973–4	NW Mediterranean	88	268
1966–8	Baltic	30	30
1965–8	North Sea, mouth of Rhine	100–250	600–1100
1972	North Sea	13–25	30–95
1970	Canadian Atlantic coast	20	140

(Adapted from Marchand *et al.* (1976).)

substances in seafood have required special consideration; examples are Hg, Cd, Pb, As, and polychlorinated biphenyls (PCBs) and polynuclear aromatic hydrocarbons among the organics. Some representative figures from coastal waters around the U.K. are given in tables 1 and 2. Such substances may be taken in and accumulated from seawater, food organisms, or seabed sediments to levels as high as 10^4–10^6 times above ambient. Such bioaccumulation is rarely seriously harmful to the marine animals or plants but the levels reached may be unacceptable in seafood, particularly in countries where regulations regarding food quality are based on somewhat arbitrarily chosen numbers (the so-called 'maximum acceptable levels') rather than on an understanding of total intake from all sources of potentially damaging substances. For example, a maximum acceptable level of mercury of 0.5 mg/kg in fish has been recommended for worldwide use, even in countries with a low consumption per person. However, in the U.K., the occurrence of a small proportion of landings with levels substantially exceeding 1 mg/kg has been judged (M.A.F.F. 1971, 1973a) to pose no health hazard to the consumer. Much of the inorganic mercury discharged to the marine environment is slowly converted to the much more toxic methylmercury and in Japan at Minimata and elsewhere there were fatalities and many permanent disabilities resulting from consumption of seafood contaminated by mercury discharged in industrial waste. In several countries restrictions on fishing, e.g. in estuaries and lakes, have been imposed because of high mercury content of fish.

Oceanic pelagic fish such as tuna, swordfish, marlin and sharks may have mercury contents of several milligrams per kilogram wet mass in the large old fish but these are 'natural' and are not derived from industrial pollution. Levels in museum specimens caught a century ago have been found to be similar. Nevertheless, in the U.S.A. and Canada fisheries for swordfish, some tuna and projected developments of fisheries for sharks and dogfish have been 'killed' because of high mercury contents. The main sources of oceanic mercury appear to be atmospheric transport of mercury lost from the land by degassing or derived from volcanic activity on shore, and undersea additions from geothermal activity especially in ocean bed spreading areas. Upwelling deep water, rich in nutrients, on which the production of some of the World's greatest fisheries depends, is likely to be richer in mercury than the surface water it displaces owing to additions from undersea geothermal activity. This seems to be the main pathway by which high levels in oceanic tuna, swordfish, marlin, etc., are maintained. Black marlin (*Makaira indica*) caught off Queensland (this species is eaten in Japan) may average 7.3 mg/kg wet mass (maximum 165 mg/kg) with levels in the liver up to 63 mg/kg wet mass (Mackay *et al.* 1975). These fish are also high in selenium which presumably exerts a protective effect against damage by mercury. A similar situation exists in seals in the southern North Sea (Koeman *et al.* 1975).

The occurrence of raised levels of mercury locally in estuarine and coastal fish and shellfish can be prevented only by stricter regulation of the use of mercury and of its discharge in industrial waste. Unfortunately, at local coastal 'hot spots' resulting from industrial contamination, the mercury is present in sediments either closely bound to organic matter or as insoluble sulphide in the deeper layers, and is only slowly released to interstitial and overlying waters. Natural reduction of these areas of contaminated sediment is very slow and much of the mercury discharged appears likely to be in organic form. Levels of mercury in fish and shellfish in such areas, e.g. off the Mersey, appear unlikely to fall significantly within a decade even though the main flux from land sources has been substantially reduced.

Cadmium has been found in edible shellfish (molluscs and crustaceans) from the U.K. at levels up to 49 mg/kg wet mass (M.A.F.F. 1973b). Even higher levels were found in non-commercial molluscs such as limpets and dog whelks taken from the Bristol Channel in areas probably heavily contaminated by effluent from a zinc smelting works. However, levels in shrimps, cockles and mussels from the nearest commercial fisheries in South Wales and along the Somerset coast did not differ materially from those recorded from other fishing areas. Very high levels in the brown meat of edible crabs were found in specimens taken off the southwest of England and Shetland outside the influence of industrial pollution. These are natural levels which are unlikely to have varied over the last 100 years. Having regard to the low average consumption of shellfish in the U.K. (1.5 g/day), the high content of Cd is not considered to constitute an unacceptable health risk (M.A.F.F. 1973b); however, the Toxicity Subcommittee of the Committee on Medical Aspects of Chemicals in Food and the Environment (Department of Health and Social Security) recommended that epidemiological investigations should be made of consumers of large amounts of shellfish 'if identifiable, to determine whether their body burden of cadmium is excessive compared with the rest of the population' (M.A.F.F. 1973b). The results of any such investigation made have not been published.

Much of the lead discharged to the atmosphere by motor vehicle exhausts reaches the sea by one route or another and there has been a marked build-up of lead in shallow water muds and marine animals around the coasts of the U.K. This can be directly correlated with the introduction and use of leaded petrols. Existing levels in filter-feeding bivalve molluscs such as mussels and oysters are uncomfortably high (up to 8.4 mg/kg wet mass (M.A.F.F. 1975)) and there has been a steady spread out from the coast in the area affected. The Pharmacology Subcommittee (D.H.S.S.) concluded (in the M.A.F.F. (1972) Survey Report) that 'in view of the low consumption of shellfish by the average individual in the United Kingdom, this [the local existence of lead values 5–6 times those found in shellfish generally] was not considered to constitute a health hazard to the majority of the population'. However, the Subcommittee noted that 'a high dietary consumption of shellfish with high lead content could conceivably be a risk for selected individual consumers', and recommended that epidemiological investigations should be undertaken on such consumers. The Supplementary Report (M.A.F.F. 1975) further considers the position of shellfish but makes no special recommendation. The weekly intake of lead in food and drink by the average person in the U.K. is estimated to be 1.2 mg against a provisional tolerable level of 3 mg established by the Joint F.A.O./W.H.O. Expert Committee on Food Additives in their 16th Report. The results of 4621 analyses of shellfish are given in the M.A.F.F. Report; the highest mean contents (0.5–4.0 mg/kg wet mass) were from shellfish taken between Hartland Point (Devon) and St David's Head (Pembroke). Values in fish (2650 samples) reached 6.8 mg/kg wet mass (see note on p. 29), but weighted means were all below 1.0 mg/kg wet mass.

By no means all of this lead comes from petrol (old mine workings, smelters, fresh water supplies and sewage are important sources), but a decrease in the total amount of lead discharged from motor exhausts would be welcomed by those concerned with the welfare of inshore fishermen.

Fantastic amounts of copper and zinc are accumulated by marine animals, particularly filter feeders such as bivalve molluscs, but without apparent harm to themselves or to the consumers of seafood. Oysters may contain so much copper in areas affected by drainage from old mine workings, such as off the mouth of Restronguet Creek in the Fal estuary, that they become green and unsaleable and have to be relaid for a summer in relatively copper-free water to get rid of the excess and become white. Nereid worms from estuaries in the southwest of Britain have been found containing up to 0.1 % of Cu and 0.035 % of Zn in their dry matter. A substantial degree of adaptation to these high metal levels is evident in these worms (Bryan 1977).

It has been known for a long time that arsenic is present in substantial quantities in the sea and may be accumulated by marine animals used as food, e.g. oysters (Chapman & Linden 1926). Recent investigations have shown, however, that greater concentrations may be present in brown seaweed, squids, crustaceans and gastropod molluscs than occur in bivalves (Leatherland & Burton 1974; Bryan 1977). In the Canadian Pacific crab, *Cancer magister*, up to 37.8 mg/kg wet mass have been found (Le Blanc & Jackson 1973); comparable analyses from the European crab, *Cancer pagurus*, do not seem to have been published. Much of the arsenic is present in organic form; it has been suggested that the end point of the organic cycle may be dimethyl arsinic acid (Wood 1974), which is much less toxic than inorganic arsenite, but others (Edmonds *et al.* 1977) have demonstrated the presence of arsenobetaine in shellfish. It appears that marine organisms may have developed a detoxification process protecting them from the effects of arsenic. Although further investigations of arsenic levels in food generally are now being made, it seems unlikely that they will reveal any problems connected particularly with the consumption of seafood in Britain. Nevertheless, the judgement of Gesamp (1976) may be appropriately quoted: '. . . the richest sources of dietary arsenic appear to be seafoods, especially shellfish and crustaceans; severe or prolonged pollution of the sea with arsenic must be avoided in those areas from which these species are taken, especially if seafoods form a substantial proportion of the normal diet of the population.' The context here is worldwide and it should be recalled that in Japan, for example, not only is the average consumption of seafood about $3\frac{1}{2}$ times that in Britain but the range of species consumed is much wider and includes many that are cultivated in shallow coastal waters and estuaries.

Although persistent organohalogen substances, such as organochlorine pesticide residues and PCBs, are found in marine animals on a more or less worldwide scale, and may be accumulated, particularly perhaps by marine plankton, there is no evidence that they have caused more than local damage, e.g. in estuaries affected by heavy factory discharges or through accidental spills, with three important exceptions. First, it is well established that eggshell thinning, resulting in low reproductive success, has occurred in several species of marine fish-eating birds (Ratcliff 1970; Gress *et al.* 1973). This is attributed to accumulation of organo-chlorine pesticide residues, particularly DDE, a breakdown product of DDT, and related compounds. Secondly, periodic unexplained mortalities of seabirds have occurred (some around the U.K.) and those examined have been found to contain elevated levels of PCBs, for example, but no direct cause–effect relation has been established (N.E.R.C. 1971). Thirdly, seal losses and reproduction failure (e.g. in the Baltic) have also been associated by some scientists with enhanced levels of DDT and PCBs (Helle *et al.* 1976; Helle 1976). Similar effects have been observed in mink fed with PCBs. However, there is no evidence that the presence of residues of these substances in seafood constitutes a human health hazard. Perhaps an exception should be made here for fish liver oils from the Baltic Sea, which have been withdrawn from the human consumption market in some countries.

Some organochlorine pesticides have been in use for more than 30 years but PCBs have been employed for an even longer period. It is remarkable that PCBs were not discovered in the marine environment until 1966 (Jensen 1966), being previously included with organochlorine pesticides in analytical procedures; at that time their annual production had reached hundreds of thousands of tonnes. During the last decade their use has declined, being confined, in theory at least, to situations where release to the environment is not expected. PCBs are, appar-ently, even more widespread and persistent that DDE; their half-life in the marine environment is not known. Total production and usage of the most commonly used organochlorine pesticides, DDT, BHC, dieldrin, endosulfan, endrin and aldrin, has almost certainly not declined on a worldwide basis; a diminution in their employment in the developed countries of the Northern Hemisphere has been more than balanced by an increase in developing countries for crop pro-tection, animal pest control and public health purposes. Since they are widely dispersed through atmospheric transport it seems likely that levels in the marine environment will remain elevated for several decades. Provided that accidental spills and the discharge of heavily contaminated wastes from pesticide manu-facturers and large industrial usage (e.g. in mothproofing and timber treatment) can be avoided, there is no reason to anticipate any progressive increase in risk to marine ecosystems. However, the combination of very high initial toxicity and persistence is a dangerous one and vigilance in the use of organochlorine pesticides and similar substances needs to be maintained.

Generally chlorinated organic substances of lower molecular mass seem to be both less toxic to marine animals and less persistent in the marine environment. Data are sparse, but Jensen *et al.* (1972) have provided information regarding the presence of waste products from vinyl chloride production in the North Atlantic, and Murray & Riley (1973) give concentrations of $CHCl_3$, CCl_4, $CHCl{=}CCl_2$ and $CCl_2{=}CCl_2$ in northwestern Atlantic surface waters. Some information has been provided on the possible effect of discharge of vinyl chloride wastes at sea (which is now generally discouraged) but the biological significance of the chlorinated hydrocarbons of lower molecular mass commonly detected in the sea is not well documented (N.A.S. 1975).

OIL

Oil pollution of the sea has received a great deal of attention from research workers throughout the world since the wreck of the *Torrey Canyon*. Although certain constituents of oil, particularly the aromatic hydrocarbons, can be shown in laboratory experiments to be harmful to marine organisms (sometimes at concentrations of less than 1 part/10^6), no evidence has been produced to show that floating oil has ever affected the recruitment to any fish or shellfish stock (Gesamp 1977). The destruction by oiling of seabirds is an intractable problem and some reduction of populations of certain auks (guillemots, razorbills and puffins) is already evident, e.g. in SW England. Other groups, e.g. certain sea ducks, divers and grebes, are also threatened and the prospects of reducing the damage seem to be very poor (Bourne 1977). The oiling of beaches, the inconvenience caused and the high cost of cleaning are too familiar to need discussion here.

Concern about health risks associated with oil has centred upon the presence in seafood of polynuclear aromatic hydrocarbons, including known carcinogens such as benzo(a)pyrene and benz(a)anthracene, and the suggestion that these are accumulated by fish and shellfish to such an extent as to constitute a health risk when the seafood is eaten. Although such substances are taken in and accumulated by fish and shellfish they are also quickly discharged when the animals are placed in oil-free water. Fish and crustacea also metabolize polynuclear aromatic hydrocarbons but the ability of bivalve molluscs to do so is doubtful and their discharge rates are slower. This subject has excited acute controversy, particularly in the U.S.A., but it is generally agreed that oil is not the major source of such polynuclear aromatic hydrocarbons in marine animals: other land sources are more important and the substances are also synthesized by marine biota. In view of the ubiquitous occurrence of BaP in the human environment and its presence in most foods, it must be concluded that the additional risk from seafood contaminated with oil, if there is one, is insignificant in relation to those to which we are exposed in daily life (motor exhausts, coal and wood fires, garden bonfires, smoked foods, leaf vegetables, etc.).

Evidence is accumulating, in the United States, of abnormal conditions in fish and invertebrates, particularly bivalve molluscs such as clams and oysters, which conditions some workers have described as neoplasms; these are said to be particularly evident in situations of chronic pollution and areas of serious oil spills (Barry & Yevich 1975; Yevich & Barszcz 1976; Brown *et al.* 1976; Gesamp 1977). No firm connection between oil hydrocarbons and these conditions seems to have been established and they have not been produced experimentally in the laboratory by exposure to oil. Work on these abnormal conditions is, however, hampered by lack of understanding of the histopathology of marine invertebrates (and even of normal histology) and by the absence of records from populations not subject to obvious pollution stress.

Sewage

Sewage and other high-b.o.d. wastes almost certainly constitute the most serious marine pollution problem. Although their contained nutrients contribute materially to production in the sea, there is increasing evidence to suggest that this has become unbalanced in such heavily affected areas as the southern North Sea. The connection, if any, between excessive sewage enrichment and recurrent blooms of highly poisonous dinoflagellates (such as that which caused over 80 cases of illness on the northeast coast of England in 1968 and the institution of annual monitoring of toxin levels in mussels from that area) has not been established, but certain nutrients are known to be required before such a bloom is triggered off. Direct transmission of typhoid, cholera and viral hepatitis by shellfish eaten raw or lightly cooked is well established and although purification and heat sterilization techniques have been worked out they add materially to the cost of production. There is continuing controversy over the hazards of bathing on sewage-contaminated beaches. The official view in the U.K. is that there is no proof that bathers run any additional risks except where pollution is so gross as to be repulsive on aesthetic grounds. In warmer countries (particularly on the Mediterranean) some scepticism is evident but has not so far been matched by epidemiological evidence or much improved sewage disposal practice in popular tourist areas.

From the point of view of those concerned with the living resources of the sea, deoxygenation in estuaries and coastal waters, and the deposit of organic-rich sediment and its substantial load of metals and persistent organics, are the principal adverse characteristics of sewage. On a worldwide basis well over 90% of sewage from coastal areas enters the sea untreated. The position in the U.K. is little better and the policy of extending outfalls to avoid beach contamination results in the inclusion of industrial wastes that might otherwise be unacceptable and possibly the direct discharge of toxic materials on to nursery grounds for young fish or productive shellfish areas. If sewage is fully treated there is a heavy residue of sludge and in coastal areas there is increasing pressure to dump this at sea. It is generally unwanted by farmers because of its content of toxic substances and it may be unwelcome in the sea for the same reason. The only solution to this dilemma is to exclude certain metals and persistent substances from waste discharges at their source, by providing treatment at the industrial plants producing them.

DISCUSSION AND CONCLUSIONS

Because of the breadth and multidisciplinary nature of the subject, only a few aspects of marine pollution can be considered in a communication of reasonable length. This paper is concerned only with damage to living resources, widely interpreted, and human health hazards. Only the latter is strictly relevant to the purpose of the Discussion Meeting, the study of long-term toxic effects in man. However, some lessons may be learned from what has been observed in the coastal seas and oceans. While it is to be expected that oil hydrocarbons will be found worldwide, with some concentration near the main shipping routes, it is something of a surprise to find that man-made chlorinated organic substances, such as DDT and its associated products and PCBs, can be found wherever they are looked for in marine materials. To some extent the sensitive analytical apparatus makes the problem but there is no doubt that within 30–40 years these complex organochlorine substances have dispersed over the entire ocean. There is equally little doubt that aerial transport has been the main agency of spread. It has been suggested that if manufacture of unnatural substances ceased tomorrow, within 20 years we should have difficulty in finding them even with the most sensitive analytical equipment. We do not know if this is so but it would be optimistic to expect any material decrease within a decade during the present phasing out of the use of some of the substances to which objections have been most strongly raised on environmental grounds. Nevertheless, it appears that analyses of marine species off northeast Europe have shown a decline in levels of DDT and its breakdown products during the last 5 years (H. Roderick, personal communication). Fortunately the presence of pesticide residues and PCBs in seafood has not been shown to give rise to any human health hazard, but had the reverse been the situation it would have been impossible to have reduced quickly the content in fish and shellfish of these man-made organochlorine substances by merely shutting off the sources of loss to the sea. It would have been necessary also to ban the sale of contaminated seafood possibly for periods of up to 10 years. Some form of early warning system directed against the introduction and general use of substances known to be both resistant to breakdown and highly poisonous to fish and shellfish is highly desirable. Routine toxicity testing, as practised, has many defects and the necessary procedures for effective examination of 'new' substances are both costly and time-consuming. Routine tests sometimes include fish, and farm-bred rainbow trout are generally employed because of convenience and reasonable standardization, but it seems rather reckless to assume, for example, that hazards to shrimps can be measured by the reactions of trout. Since the sea is so much more extensive than the land and the effects of introductions of persistent pollutants there are mostly irretrievable, there is certainly a case for introducing at least one marine species of value as food into routine screening of substances for environmental use.

Although it has been suggested (Halstead 1972) that heavy pollution of tropical waters may trigger off abnormal production cycles which cause normally accept- able fish to become poisonous, the evidence for such a belief appears to be weak. Nor is there any clearcut evidence linking pollution with blooms of poisonous dinoflagellates although these seem to be occurring more frequently than formerly and in areas, such as the North Sea, not historically associated with such phenom- ena. The most that can be said at present is that an imbalance of nutrients associated with sewage enrichment could play a part in producing these algal blooms. Not all of them are poisonous but they may nevertheless be objectionable – the term 'hypertrophication' is becoming common in marine biological literature and usually implies unsatisfactory conditions.

In the context of human health hazards from seafood, the metals mercury, lead and cadmium are the most important pollutants, when the risks of the spread of enteric diseases and viral hepatitis from eating sewage-contaminated shellfish have been set aside. In respect of the latter, since it is not possible to visualize a situation where a large part of the world's sewage does not reach the sea, the hope of controlling the spread of these diseases by seafood rests upon identifi- cation of shellfish beds at risk and the use of purification techniques. These measures must, however, be coupled with the designation of special areas for shellfish production and their protection from heavy pollution. The answer is not to pollute and purify but to protect and still purify to the necessary degree as a further preventive measure. The point must, perhaps, be made that although epidemiological evidence fails to show any connection between the consumption of purified shellfish and viral hepatitis, there is no experimental evidence yet that viral infective agents can be removed from shellfish by present methods of puri- fication. For a general review of the subject of shellfish hygiene see Wood (1976).

Returning to the potentially harmful metallic residues, it is the total intake from all sources in the diet that is relevant. An occasional meal of fish or shellfish from a metal-contaminated coastal area (a local 'hot spot') will scarcely influence the total intake level. In countries such as Britain, with a generally modest average consumption of fish and particularly of shellfish, there seems to be no possibility of seriously elevated intakes of mercury, lead or cadmium from sea- food by the general population. However, surveys of local habits, conducted in connection with the control of exposure to radioactive waste discharged to the sea, have shown that a small minority with ready access to fish and/or shellfish may consume unexpectedly large amounts and may do so regularly. Such very limited investigations as have been made of abnormally heavy seafood consumers in the United Kingdom to investigate possible harm from metal intake in the diet have not, however, revealed any cause for concern.

Estuaries and coastal waters where the sediments have become heavily contaminated with metals such as mercury, lead and cadmium (the 'hot spots' referred to above) present a continuing problem long after the source of contamination has been stopped or reduced. Sources of mercury or cadmium can be identified on land, in industrial usage, but the principal source of lead may be atmospheric transport of petrol additives. Stoppage at source is not, however, quickly followed by diminished levels in fish and shellfish frequenting the contaminated bottom. Metals are closely bound in sediments and may well become stabilized in anoxic mud as insoluble sulphides. This seems to happen with mercury and the small quantities released into interstitial waters in the sediments are likely to prove a potent source of methylmercury and one that may persist for decades. To get rid of this mercury trapped in sediments it may be necessary to dredge up the contaminated deposits and to disperse them widely in deep water offshore. Such action would at present be in contravention of the international conventions (London and Oslo) controlling dumping at sea.

Although cadmium is held in sediments and some local areas of concentration have been found, e.g. in the Bristol Channel, a high dietary intake of cadmium seems most likely to arise from exceptionally heavy consumption of the brown meat of crabs. Their high content of cadmium is not derived from pollution. Although average consumers of fish and shellfish are not at risk in Britain and are never likely to be, because of cadmium, an addiction to crab paste (mainly brown meat) combined with heavy smoking (shown to increase the body burden of cadmium) could be a recipe for trouble since cadmium accumulates in the body throughout life. Production of cadmium within the countries of the European Economic Community (E.E.C.) rose by $ca.$ 50 % between 1967 (2246 t) and 1972 (3415 t).

World production of lead is on a much larger scale than that of mercury or cadmium. Of a total production of about 3.5 Mt, approximately 10 % is used in leaded petrols. Although the permitted level of lead additives has been reduced in environmentally conscious countries, the total number of motor vehicles in use continues to rise. For some time at least the worldwide level of emissions seems likely to remain about the same. Lead is persistent in the marine environment and the marine sediment compartment probably contains the largest amounts. Levels in coastal sediments around Britain do not seem to have changed much in recent years and something approaching an equilibrium situation may have been reached. Areas of elevated lead content due to industrial contamination are not difficult to find. They give rise to uncomfortably high levels in bivalve shellfish, in particular, which cause some concern because of continuing uncertainty about the effects of exposure to low levels of lead on human health and mental capacity. Although further restrictions on the use of lead additives in petrol may be introduced and lead may be largely replaced by plastics in domestic plumbing, consequential changes in the flux of lead to the sea will take several decades to have an appreciable effect on the lead content of seafood. Bivalve molluscs will continue to accumulate lead from the reserves in the sediments. In contrast to those contaminated by mercury, where 'hot spots' are local and usually limited in area and possibly reducible by dredging, lead-contaminated sediments occur widely and will only slowly lose lead by natural processes of release and dispersion. Their decontamination by any other means is out of the question.

Suggestions that the productivity of the sea may be severely damaged by pollutants which are slowly building up to dangerous levels appears to be without support from factual evidence. Generally, production has been well maintained even in those semi-enclosed seas receiving notably high discharges of sewage and industrial waste. However, it is clear that those problems which have been allowed to develop from the accumulation of persistent and potentially dangerous substances are here to stay, principally because of accumulation and stabilization in estuarine and coastal sediments. Such accumulations will only be very slowly reduced, over decades, by natural processes when the flux of pollutants from the land and the air has been halted.

Note added in proof, 16 *May* 1979 (see p. 22). Recent work has indicated that lead values may be artificially increased by contamination during sampling and analysis.

REFERENCES (Cole)

Barry, M. M. & Yevich, P. P. 1975 The ecological, chemical and histopathological evaluation of an oil spill site. *Mar. Pollut. Bull.* **6**, 171–173.

Bourne, W. R. P. 1977 Seabirds and pollution. In *Marine pollution* (ed. R. Johnston), pp. 404–502. London, New York and San Francisco: Academic Press.

Boyden, C. R. 1975 Distribution of some trace metals in Poole Harbour, Dorset. *Mar. Pollut. Bull.* **6**, 180–187.

Brown, R. S., Wolke, R. E. & Saila, S. B. 1976 A preliminary report on neoplasia in feral populations of the soft-shell clam *Mya arenaria*: prevalence, histopathology and diagnosis. In *Proc. First Internat. Colloq. Invert. Pathol., Kingston, Ontario, Canada*, pp. 151–159.

Bryan, G. W. 1977 Heavy metal contamination in the sea. In *Marine pollution* (ed. R. Johnston), pp. 185–302. London, New York and San Francisco: Academic Press.

Chapman, A. C. & Linden, H. 1926 On the presence of compounds of arsenic in marine crustaceans and shell fish. *Analyst* **51**, 563–564.

Cole, H. A. & Holden, M. J. 1973 History of the North Sea fisheries 1950–1968. In *North Sea science* (ed. F. D. Goldberg), pp. Cambridge, Mass.: M.I.T. Press.

Edmonds, J. S., Francesconi, K. A., Cannon, J. R., Raston, C. L., Skelton, B. W. & White, A. H. 1977 Isolation, crystal structure and synthesis of arsenobetaine, the arsenical constituent of the western rock lobster *Panulirus longipes cygnus* George. *Tetrahedron Lett.*, pp. 1543–1546.

Gesamp 1973 *Report of the Fifth Session.* Gesamp V/10, October 1973, 87 pages.

Gesamp 1976 *Review of harmful substances. Reports and studies* no. 2, 80 pages.

Gesamp 1977 *Impact of oil on the marine environment. Reports and studies* no. 6, 250 pages.

Gress, F., Risebrough, R. W., Anderson, D. W., Kiff, L. F. & Jehl, J. R. 1973 Reproductive failures of double-crested cormorants in southern California and Baja California. *Wilson Bull.* **85**, 197–208.

Halstead, B. W. 1972 Toxicity of marine organisms caused by pollutants. In *Marine pollution and sea life* (ed. M. Ruivo), pp. 584–594. London: Fishing News (Books).

Helle, E. 1976 PCB levels correlated with pathological changes in seal uteri. *Ambio* **5**, 261–263.

Helle, E., Olsson, M. & Jensen, S. 1976 DDT and PCB levels and reproduction in Ringed Seal from the Botham Bay. *Ambio* **5**, 188–189.

Helliwell, P. R. & Bossanyi, J. (eds) 1975 *Pollution criteria for estuaries.* London: Pentech Press.

I.C.E.S. 1974 Report of working group for the international study of the pollution of the North Sea and its effects on living resources and their exploitation. *Cooperative Research Report* no. 39, 191 pages.

Jensen, S. 1966 Report of a new chemical hazard. *New Scient.* **32**, 612.

Jensen, S., Jernelov, A., Lange, R. & Palmork, K. H. 1972 Chlorinated by-products from vinyl chloride production: a new source of marine pollution. In *Marine pollution and sea life* (ed. M. Ruivo), pp. 242–244. London: Fishing News (Books).

Jensen, S., Johnels, M., Olsson, M. & Otterlind, G. 1969 DDT and PCB in marine animals from Swedish waters. *Nature, Lond.* **224**, 247–250.

Key, D. 1972 *The Stanswood Bay oyster fishery. M.A.F.F. Shellfish Information Leaflet* no. 25. (14 pages.)

Key, D. 1977 Natural stocks of oysters on public grounds in the Solent and adjoining harbours in 1976. *M.A.F.F. Fisheries Notice* no. 50. (17 pages.)

Koeman, J. H., Van de Ven, W. S. M., de Goeij, J. J. M., Tijioe, P. S. & van Haaften, J. L. 1975 Mercury and selenium in marine mammals and birds. *Sci. tot. Environ.* **3**, 279–287.

Leatherland, T. M. & Burton, J. D. 1974 The occurrence of some trace metals in coastal organisms with particular reference to the Solent region. *J. mar. biol. Ass. U.K.* **54**, 457–468.

Le Blanc, P. J. & Jackson, A. L. 1973 Arsenic in marine fish and invertebrates. *Mar. Pollut. Bull.* **4**, 88–90.

Mackay, N. J., Kazacos, M. N., Williams, R. J. & Leedow, M. I. 1975 Selenium and heavy metals in Black Marlin. *Mar. Pollut. Bull.* **6**, 57–60.

M.A.F.F. 1971 *Survey of mercury in food.* (33 pages.) London: H.M.S.O.

M.A.F.F. 1972 *Survey of lead in food.* (31 pages.) London: H.M.S.O.

M.A.F.F. 1973a *Survey of mercury in food. Supplementary Report.* (34 pages.) London: H.M.S.O.

M.A.F.F. 1973b *Survey of cadmium in food.* (31 pages.) London: H.M.S.O.

M.A.F.F. 1975 *Survey of lead in food: first supplementary report.* (34 pages.) London: H.M.S.O.

Marchand, M., Vas, D. & Duursma, E. K. 1976 Levels of PCB's and DDT's in mussels from the N.W. Mediterranean. *Mar. Pollut. Bull.* **9**, 65–68.

Murray, A. J. & Riley, J. P. 1973 Occurrence of some aliphatic hydrocarbons in the environment. *Nature, Lond.* **242**, 37–38.

N.A.S. 1975 *Assessing potential ocean pollutants.* (438 pages.) Washington, D.C.: National Academy of Sciences.

N.E.R.C. 1971 *The seabird wreck in the Irish Sea, Autumn 1969.* Publication Series C, no. 4, 17 pages.

Ratcliff, D. A. 1970 Changes attributable to pesticides in egg breaking and eggshell thickness in some British birds. *J. appl. Ecol.* **7**, 67–115.

Ruivo, M. (ed.) 1972 *Marine pollution and sea life.* London: Fishing News (Books).

Wood, J. M. 1974 Biological cycles for toxic elements in the environment. *Science, N.Y.* **183**, 1049–1052.

Wood, P. C. 1976 *Guide to shellfish hygiene. W.H.O. Offset Publication* no. 31. (80 pages.)

Yevich, P. P. & Barszcz, C. A. 1976 Gonadal and haematopoietic neoplasms in *Mya arenaria. Mar. Rev.* **38** (10), 42–43.

Proc. R. Soc. Lond. B. **205**, 31–45 (1979)

Printed in Great Britain

Hazards to wintering geese and other wildlife from the use of dieldrin, chlorfenvinphos and carbophenothion as wheat seed treatments

By P. I. Stanley and P. J. Bunyan

Pest Infestation Control Laboratory, Ministry of Agriculture, Fisheries and Food, Hook Rise South, Tolworth, Surbiton KT6 7NF, U.K.

Chemical treatments of cereal seeds are used in the United Kingdom to prevent damage by a number of pests including the wheat bulb fly, which is a serious pest of winter wheat. The persistent organochlorine dieldrin was introduced in the 1950s as a seed treatment but caused the death of large numbers of grain eating birds and gave rise to unacceptable environmental contamination. The withdrawal of dieldrin as a seed treatment was made possible by the introduction of two less persistent organophosphate insecticides, chlorfenvinphos and carbophenothion. Although the introduction of these chemicals has been beneficial in reducing environmental contamination, some side-effects on wildlife have still been discernible and carbophenothion has now been withdrawn from use in Scotland owing to the deaths of wintering geese from carbophenothion poisoning. Subsequent laboratory studies have demonstrated that *Anser* geese are particularly susceptible to carbophenothion poisoning, and the underlying biochemical mechanism has been investigated. The fundamental problem of species variation in toxicity among the organophosphorus and carbamate pesticides which this investigation illustrates presents difficulties for registration authorities when they are considered for clearance for agricultural use. The implications of the environmental problems encountered with dieldrin, chlorfenvinphos and carbophenothion for the pre-clearance testing of new chemicals are discussed and the critical surveillance of the early years of commercial use of a chemical is recommended to support pre-clearance studies aimed at assessing the potential hazard to the environment.

Introduction

During the past 30 years, a vast range of novel organic compounds has been synthesized and developed for use in industry, agriculture, medicine and other areas of man's activities. It has been estimated that over 100 000 chemicals are in use in British industry and that a large number of new industrial chemicals is introduced each year (Health and Safety Commission 1977). The requirement for greater agricultural yields has stimulated the development of a large range of pesticides. Insecticides account for only a small part of the total agricultural chemical usage but the use of organochlorine insecticides has led to serious environmental problems that have had a profound influence on the approach to the

[31]

use of chemicals by man. The first indication of these problems came from wild-life studies and in particular from the observed decline of populations of predatory birds. Later research revealed the ubiquitous contamination of the environment by the persistent organochlorine residues and showed that this could have subtle effects on ecosystems. The inherent problems in the use of persistent agricultural chemicals are now accepted and the environmental problems that have arisen from the industrial use of the polychlorinated biphenyls have demonstrated that industrial chemicals can also be released into the environment on a large scale.

The recognition of the potential hazards from the use of novel compounds has led to the establishment in many countries of registration authorities to control the use of chemicals in agriculture and industry. A major feature of the strategy employed by registration authorities in recent years has been the phasing out of the persistent chemicals and their replacement by less persistent alternatives. This move has already resulted in a reduction in the level of environmental contamination which is reflected in the recovery of some predatory bird populations. However, recent experience in the United Kingdom with two less persistent insecticides suggests that although they do not cause environmental contamination, they may present other hazards to wildlife and the environment.

Winter wheat is liable to attack by the wheat bulb fly (*Delia coarctata*), and the insecticide dieldrin (1,2,3,4,10,10-hexachloro-6,7-epoxy-1,4,4a,5,6,7,8,8a-octa-hydro-*exo*-1,4-*endo*-5,8-dimethanonaphthalene) was introduced in the 1950s as a wheat seed treatment. This use of dieldrin led to the deaths of large numbers of granivorous birds and to the contamination of predatory species. Dieldrin has now been withdrawn as a cereal seed treatment and has largely been replaced by two organophosphate insecticides, chlorfenvinphos (2-chloro-1-(2,4-dichloro-phenyl)vinyl diethyl phosphate) and carbophenothion (*S*-(4-chlorophenylthio-methyl)diethyl phosphorothiolothionate).

The Pest Infestation Control Laboratory monitors the introduction of new pesticides for effects on wildlife and investigates reports of deaths which are suspected to have been caused by agricultural chemicals (Pest Infestation Control 1973; Pest Infestation Control Laboratory 1975, 1978). As a result of these investigations, carbophenothion has been shown to present a serious hazard to wintering geese and its use as a cereal seed treatment has now been curtailed in the important geese wintering areas in Scotland. The case history discussed in this paper of the withdrawal of dieldrin and the introduction of chlorfenvinphos and carbophenothion and the subsequent wildlife problems has many implications for the testing of new chemicals for environmental hazard and emphasizes the need for the critical surveillance of the early years of use.

THE WHEAT BULB FLY AS A CEREAL PEST AND THE NEED FOR CHEMICAL CONTROL

The wheat bulb fly is probably the most serious insect pest of winter wheat in Great Britain (M.A.F.F. 1973). The areas most subject to serious attack are those in Scotland (Angus, Fife and Lothian) and the eastern counties of England. It has been estimated that the mean annual loss due to this insect, if no insecticide were used, would be equivalent to the yield from 16000 ha of winter wheat (Graham 1977). The larvae emerge in mid-January and enter the plant below ground at the first node. Sowing the seed early and with adequate fertilizer so that the plant is well developed by the time the larvae emerge reduces damage but effective and economic control can only be attained by using an insecticide seed treatment. Alternative control techniques such as the use of granular formulations of insecticide for incorporation into the soil at the time of sowing or spray application of insecticide at the time of attack are considerably more expensive and involve additional farming operations.

Farmers are advised to sow treated seed at a shallow depth so that the region of the plant which is vulnerable to attack between the seed and the surface is as short as possible and is within the zone where the insecticide on the seed is effective. This advice increases the hazard presented to wildlife.

HAZARDS TO WILDLIFE FROM THE USE OF DIELDRIN AS A WHEAT SEED TREATMENT

During the 1950s the cyclodiene insecticides aldrin (1,2,3,4,10,10-hexachloro-1, 4, 4a, 5, 8, 8a-hexahydro-*exo*-1, 4-*endo*-5, 8-dimethanonaphthalene), dieldrin and heptachlor (1, 4, 5, 6, 7, 8, 8-heptachloro-3a, 4, 7, 7a-tetrahydro-4, 7-methanoindene) were introduced as cereal seed treatments. As early as the spring of 1956, reports were received of the deaths of large numbers of seed-eating birds after the use of the cyclodiene seed treatments (Turtle *et al.* 1963). The scale of the wildlife deaths in 1960 provoked widespread concern and measures were introduced, to take effect on 1 January 1962, to limit their use to autumn sowings when there was a real danger of attack by wheat bulb fly (Sanders 1961). Cramp *et al.* (1963) reported that these restrictions were effective in reducing the incidence of bird deaths. Mainly as a result of residues found in meat and vegetables, the Cook review (1964) recommended further limitations on the use of the cyclodiene insecticides particularly in dips and sprays for sheep and in fertilizer mixtures. By 1967, the agriculture uses of the cyclodiene insecticides were limited mainly to seed treatments for autumn sown winter wheat and sugar beet. In 1969, the *Further review of certain persistent organochlorine pesticides used in Great Britain* recommended that the remaining agricultural uses of the cyclodiene pesticides should be continuously assessed with a view to withdrawal as soon as satisfactory non-persistent alternatives were available.

TABLE 1. WILDLIFE INCIDENTS ATTRIBUTED TO THE USE OF DIELDRIN AS A WHEAT SEED TREATMENT INVESTIGATED BY THE PEST INFESTATION CONTROL LABORATORY DURING THE PERIOD 1973–7

year	number of incidents	number of incidents attributed to dieldrin poisoning
1973	100	28
1974	113	14
1975	146	18
1976	179	9
1977	171	0

TABLE 2. SPECIES INVOLVED IN THE WILDLIFE INCIDENTS ATTRIBUTED TO THE USE OF DIELDRIN AS A WHEAT SEED TREATMENT INVESTIGATED BY THE PEST INFESTATION CONTROL LABORATORY DURING THE PERIOD 1973–7

(Figures indicate the number of incidents in which each species was involved.)

birds

feral pigeon (*Columba*)	20
woodpigeon (*Columba palumbus*)	11
pheasant (*Phasianus colchicus*)	11
kestrel (*Falco tinnunculus*)	8
short-eared owl (*Asio flammeus*)	5
rook (*Corvus frugilegus*)	3
tawny owl (*Strix aluco*)	3
barn owl (*Tyto alba*)	3
rough-legged buzzard (*Buteo lagopus*)	3
carrion crow (*Corvus corone*)	2
blackbird (*Turdus merula*)	2
robin (*Erithacus rubecula*)	2
brambling (*Fringilla montifringilla*)	2
sparrowhawk (*Accipiter nisus*)	1
goshawk (*Accipiter gentilus*)	1
collared dove (*Streptopelia decaocto*)	1
grey partridge (*Perdix perdix*)	1
jackdaw (*Corvus monedula*)	1
yellowhammer (*Emberiza citrinella*)	1
house sparrow (*Passer domesticus*)	1

mammals

cat (*Felis*)	4
fox (*Vulpes vulpes*)	2
brown rat (*Rattus norvegicus*)	2
badger (*Meles meles*)	1
dog (*Canis*)	1

Van den Heuvel (1975) reviewed the wildlife incidents investigated by the Pest Infestation Control Laboratory during the period 1963–72, and reported that even after 1967 the use of dieldrin as a wheat seed treatment presented a serious hazard to wildlife. Wildlife incidents attributed to dieldrin were occurring at the times of the autumn and spring sowing of wheat and the incidents generally involved the death of granivorous birds. The wildlife incidents attributed to dieldrin poisoning in England and Wales during the period 1973–7 are shown in table 1. During this period the coverage of the incident investigations was improved and the increase in the total number of incidents investigated does not reflect an

TABLE 3. RESIDUES OF DIELDRIN IN PINK-FOOTED GEESE
FOUND DEAD AT ABERLADY, EAST LOTHIAN

date found	dieldrin concentration (parts/10^6)†	
	pectoral muscle	liver
10 Dec. 72	7	15
22 Dec. 72	8	48
12 Jan. 73	7	26
12 Jan. 73	10	28
15 Jan. 73	7	35
15 Jan. 73	4	34

Data from Hamilton (1977).
† Residue analysis as described in Hamilton *et al.* (1976).

increase in the number of incidents occurring. The use of dieldrin wheat seed treatments was considered to be responsible for a total of 69 wildlife incidents during 1973–7. The majority of these incidents occurred in the eastern counties of England (Stanley *et al.* 1978*a*) and mainly involved granivorous species (table 2). The species principally affected was the woodpigeon (*Columba palumbus*) and the level of dieldrin in the livers of the woodpigeons found dead indicated that the birds had consumed grain treated with dieldrin and had died relatively rapidly from the acutely toxic dose (Stanley *et al.* 1978*a*). A serious outbreak of wheat bulb fly in Scotland during the late sixties led to an upsurge in the use of dieldrin seed treatments. During the spring of 1972 and 1973 large numbers of granivorous birds died in Scotland (Hamilton 1977) including a number of pink-footed geese (*Anser brachyrhynchus*) and the levels of dieldrin in specimens from Aberlady in East Lothian are shown in table 3. The pink-footed geese were thought to have consumed winter wheat seed treated with dieldrin and the residues in the liver are sufficiently large to be accepted as lethal.

During 1973 and 1974, considerable mortality of predatory birds and mammals occurred in England, indicating a serious level of environmental contamination (Stanley *et al.* 1978*a*). The unacceptable level of environmental contamination arising from the use of dieldrin and aldrin as wheat seed treatments led to a final withdrawal of their use for this purpose at the end of 1975. The withdrawal of the cyclodiene insecticides as cereal seed treatment chemicals at the end of 1975 was made possible by the introduction of the two organophosphate insecticides, chlorfenvinphos and carbophenothion, which were satisfactory alternatives and were less persistent. This withdrawal has already resulted in a sharp decline in the incidence of wildlife poisoning by dieldrin (table 1) and the level of environmental contamination should also now decline.

The toxicity of dieldrin, chlorfenvinphos and carbophenothion to birds and mammals

In the United Kingdom, new agricultural chemicals or new uses of existing chemicals are notified to the Pesticides Safety Precautions Scheme. Information is supplied by the notifier to enable the Advisory Committee on Pesticides to assess the implications for the agricultural operator, the consumer of produce and for the environment of clearing the chemical for use. The assessment of potential environmental hazard is based on laboratory toxicity and residue studies and on wildlife studies during field trials. For toxic, less persistent chemicals the most useful information is often the acute oral toxicity to a variety of species which allows the potential hazard to wildlife to be assessed.

The acute oral toxicity of dieldrin, chlorfenvinphos and carbophenothion to a range of birds and mammals is shown in table 4. Dieldrin shows a remarkably

TABLE 4. THE ACUTE ORAL TOXICITY (EXPRESSED AS L.D.$_{50}$; milligrams per kilogram) OF DIELDRIN, CHLORFENVINPHOS AND CARBOPHENOTHION TO BIRDS AND MAMMALS

(Ranges given are 95 % confidence limits.)

species	dieldrin	chlorfenvinphos	carbophenothion
rat	38–87[1–6]	9.6–39[10–12]	7–91[4, 17–20]
mouse	38–74[1, 7, 8]	117–200[12–14]	106–218[4]
rabbit	45–50[1]	300–1000[10, 12, 13]	1250[21]
dog	56–120[1, 8]	> 500[10, 13]	40[4]
mallard duck		85.5 (44.5–164)[9]	121 (95.9–152)[9]
chicken	48[8]	44–240[12, 14]	316[4]
pheasant	79 (33.3–187)[9]	107 (80–145)[15]	
pigeon	26.6 (19.2–36.9)[9]	16.4 (13.7–25.8)[15]	34.8 (31.1–38.9)[22]
quail (Coturnix)	69.7 (40.0–121)[9]	148 (121–229)[15]	56.8 (50.8–63.6)[22]
Canada goose	50–150[9]		29–35[22]
starling		3.2[16]	5.6 (3.2–10)[16]
redwinged blackbird		10[16]	7.5[16]
house sparrow	47.6 (34.3–66.0)[9]		

References: 1, F.A.O./W.H.O. (1967); 2, Gaines (1960); 3, Heath & Vandekar (1964); 4, F.A.O./W.H.O. (1973); 5, Lu et al. (1965); 6, Treon & Cleveland (1955); 7, Jolly (1954); 8, F.A.O./W.H.O. (1968); 9, Tucker & Crabtree (1970); 10, Ambrose et al. (1970); 11, Gaines (1969); 12, F.A.O./W.H.O. (1972); 13, Hutson & Hathway (1967); 14, Pickering (1965); 15, Bunyan et al. (1971); 16, Shafer (1972); 17, Edson et al. (1963); 18, Hayes (1971); 19, Shaffer & West (1960); 20, Hagan et al. (1961); 21, Buck et al. (1973); 22, Jennings et al. (1975).

consistent toxicity to all of the species tested. The acute oral toxicity is generally in the range of 30–80 mg/kg body mass and dieldrin can be regarded as acutely toxic. The single dose toxicity test does not allow assessment of the potential hazard presented by the persistence of dieldrin, but a dietary toxicity test (l.c.$_{50}$) reveals that dieldrin exhibits chronic toxicity producing mortality at a lower level in the diet than the more acutely toxic but less persistent organophosphate insecticides. The wildlife problems encountered with the use of dieldrin as a wheat seed treatment can be explained by this observation. The acute toxicity accounts for the large scale deaths of grain-eating birds and the chronic toxicity leads to the mortality observed among predators. Other effects on metabolic processes and reproduction have also been attributed to sublethal residues of dieldrin (DeWitt 1956; Atkins & Linder 1967).

The available toxicity data for chlorfenvinphos and carbophenothion demonstrate the remarkable variation in toxicity between species which is a characteristic feature of organophosphate insecticides and poses serious problems when new organophosphate insecticides are being examined through the Pesticides Safety Precautions Scheme before they come into agricultural use. For instance, observations on chlorfenvinphos administration to rat, mouse, rabbit and dog (table 4) show that toxicity between species may vary by orders of magnitude. The biochemical mechanisms underlying these differences have not generally been extensively studied with the exception of those for chlorfenvinphos in rat and dog (Hutson & Hathway 1967). They concluded that no single factor could explain the phenomenon but that the combination of a number of small differences in absorption, metabolism and excretion acting in concert could produce the observed difference in sensitivity to the compound.

Chlorfenvinphos and carbophenothion were both cleared for use as wheat seed treatments after extensive laboratory and field testing. The risk presented by the acute toxicity of the compounds to vertebrates was recognized and farmers are advised that seed dressed with either compound should not be sown after 31 December. For later sowings, a γ-HCH (1,2,3,4,5,6-hexachlorocyclohexane) seed treatment is recommended.

HAZARDS TO WILDLIFE FROM THE USE OF CHLORFENVINPHOS
AS A WHEAT SEED TREATMENT

Bunyan *et al.* (1971) investigated the toxicity of chlorfenvinphos to the pigeon, pheasant and Japanese quail (*Coturnix coturnix japonica*) and demonstrated a ninefold difference between the toxicity of chlorfenvinphos to the pigeon and to the Japanese quail (table 4). They suggested that under certain conditions, such as periods of food shortage, the sowing of chlorfenvinphos treated grain could present a risk to sensitive species such as the pigeon. This prediction has been proved to be correct by a series of wildlife incidents involving mainly feral pigeons (table 5). The birds found dead in these incidents typically exhibited severe brain cholinesterase inhibition and contained grain treated with chlorfenvinphos in their crops. Since chlorfenvinphos is used on a large scale and the incidents are relatively rare and involve an abundant species, the hazard presented by the use of chlorfenvinphos is thought to be acceptable. This is an interesting example of a potential hazard predicted on the basis of laboratory studies that has been proved to occur under field conditions.

TABLE 5. WILDLIFE INCIDENTS ATTRIBUTED TO CHLORFENVINPHOS WINTER WHEAT SEED TREATMENTS DURING THE PERIOD 1974–6

date	county	species	number of dead birds	mean chlorfen-vinphos residue in crop contents† (parts/10^6)	sample size	brain cholin-esterase inhi-bition‡
10 Mar. 74	Suffolk	woodpigeon	4	107	4	+ +
		stock dove	4	72	3	+ +
30 May 75	Fife	feral pigeon	several	34	1	n.m.§
28 Oct. 75	Norfolk	feral pigeon	23	360	1	+ +
5 Nov. 75	Leicestershire	pigeon	40	8	1	n.m.
18 Nov. 75	East Lothian	feral pigeon	7	present	—	+ +
25 Nov. 75	Lincolnshire	pigeon	20	107	1	+ +
16 Nov. 76	Lincolnshire	feral pigeon	11	10	1	+
25 Nov. 76	Lincolnshire	feral pigeon	3	10	1	+ −
26 Nov. 76	Yorkshire	feral pigeon	41	16	2	+ +

† Chlorfenvinphos residue determined by gas–liquid chromatography and identity of residue confirmed by thin-layer chromatography with esterase inhibition detection as described by Hamilton et al. (1976). The recommended treatment rate leads to a chlorfenvinphos level of approximately 550 parts/10^6.

‡ Brain cholinesterase activity measured as described by Bunyan (1973). Inhibition represented by: + +, more than 90 %; +, 50–90 %; + −, less than 50 %.

§ N.m., not measured.

HAZARDS TO WILDLIFE FROM THE USE OF CARBOPHENOTHION AS A WHEAT SEED TREATMENT

In contrast to chlorfenvinphos, the use of carbophenothion as a winter seed treatment has caused a series of incidents of ecological significance. These have involved the death of large numbers of wintering geese and are detailed in table 6. The first occurred in Scotland in 1971 and involved the death of approximately 500 greylag geese (*Anser anser*) (Bailey et al. 1972). A remarkable feature was that only geese were involved, although other granivorous birds were observed feeding in the area. This incident stimulated an investigation to ascertain whether geese are particularly susceptible to carbophenothion poisoning. Jennings et al. (1975) found that carbophenothion is equally toxic in the form of a single dose to the pigeon, Japanese quail and Canada goose (*Branta canadensis*) and that in the latter, death was accompanied by more than 90 % inhibition of brain cholinesterase and a brain residue level of carbophenothion in excess of 1 part/10^6. During the winter of 1974/5 a series of incidents (table 6) occurred in which substantial numbers of greylag and pink-footed geese died. Again no other bird or mammal species was affected. The field circumstances of these incidents have previously been described (Hamilton & Stanley 1975; Hamilton *et al.* 1976). The geese died exhibiting substantial brain cholinesterase inhibition after consuming either carbophenothion treated grain left on the surface after sowing or germinated seed

TABLE 6. DEATHS OF WINTERING GEESE ATTRIBUTED TO THE CONSUMPTION OF WINTER WHEAT TREATED WITH CARBOPHENOTHION

date	locality	species	number of dead birds[†]	reference
25 Oct. 71	Coupar Angus, Perthshire	greylag goose	500	Bailey *et al.* (1972)
21 Nov. 74	Lundie, Angus	greylag goose	325	
4 Dec. 74	Scone, Perthshire	greylag goose	46	
4 Dec. 74	Madderty, Perthshire	greylag goose	56	Hamilton & Stanley (1975)
– Dec. 74	Burrelton, Perthshire	greylag goose	24	
6 Jan. 75	Whitton Sand, Humberside	pink-footed goose	243	
– Oct. 75	St Andrews, Fife	pink-footed goose	298	Hamilton (1976)

† Based on the number of dead birds found and thus probably an underestimate of the total number of casualties.

uprooted under wet soil conditions. Brain residue levels were found to be in good agreement with those reported by Jennings *et al.* (1975).

The pink-footed geese wintering in Britain in 1975 represented approximately 85 % of the world population of this species (Ogilvie 1975). The pink-footed geese wintering in Britain breed in Iceland and Greenland. They arrive in Britain at the end of September and stay until April or early May. The greylag geese wintering in Britain in 1975 represented approximately 65 % of the world population but are a discrete population that breeds in Iceland. Greylag geese arrive in Britain in late October and stay until the second half of April. The winter distributions of both species overlap to a considerable extent with the major concentrations of each occurring in east central Scotland but with substantial numbers of pink-footed geese wintering in Lancashire.

The pink-footed and greylag geese wintering in Britain changed their feeding habits early in the present century from feeding primarily on marginal grassland to feeding almost exclusively on agricultural land. During the autumn, barley and wheat grains spilt during harvesting are the main source of food, and later harvested potato fields are frequented. When these food sources are depleted, grass and sprouting winter wheat are grazed.

Annual censuses of each species have been conducted since 1960 (Boyd &

Ogilvie 1969; Boyd & Ogilvie 1972; Ogilvie & Boyd 1976). Early estimates indicate that approximately 30000 wintering pink-footed geese were present in 1950 and the number steadily rose, reaching a peak of 89000 in 1974. The latest counts revealed approximately 71000 in 1976 (Ogilvie 1977). Greylag geese have similarly increased in number from about 28000 in 1960 to a peak of 76000 in 1973. The present wintering population (1976) is only 56000, reflecting poor breeding seasons since 1973 (Ogilvie 1977). From these population estimates it may be concluded that the number of geese killed by carbophenothion seed treatments represents a significant proportion of the world population.

The goose incidents detailed in table 6 occurred at a time when the withdrawal of dieldrin was only partly complete and thus the potential risk to the geese when dieldrin withdrawal had been completed and carbophenothion and chlorfenvinphos were used on the total acreage was considered to be unacceptable. The geographical distribution of the wintering geese allowed the population to be protected by the withdrawal of carbophenothion from use as a wheat seed treatment in Scotland. The small numbers of geese visiting eastern areas of England where wheat bulb fly treatments are used are still at risk but the farmers in these localized areas have been advised of precautions to reduce the hazard to the geese.

The apparent susceptibility of the geese to carbophenothion poisoning under field conditions could be explained by a number of factors. For instance, temperature and migration stress will influence toxicity. Nutrition, particularly protein intake, can modify the toxicity of a variety of pesticides (Shakman 1974). Geese are primarily grazers and do not possess the capacity to digest cellulose which results in a very inefficient utilization of food, a large intake and a rapid throughput (Mattocks 1971). This digestive physiology could influence the rate of absorption of carbophenothion and thus the toxicity. However, the incidents appear to have resulted from the ingestion by the geese of an acutely toxic dose of carbophenothion. From the work of Hutson & Hathway (1967), Machin *et al.* (1976), Lee & Pickering (1967) and others, it appears likely that there is an underlying biochemical mechanism to the apparent susceptibility of geese to carbophenothion poisoning.

The continuing involvement of *Anser* geese in the incidents led to a reappraisal of the testing programme of Jennings *et al.* (1975) which had involved only *Branta* geese. Species variations in toxicity are recognized for the organophosphate insecticides and a research programme was undertaken to investigate the acute and subacute toxicity of carbophenothion to the pink-footed goose, greylag goose, the Canada goose and the barnacle goose (*Branta leucopsis*) (Stanley *et al.* 1978*b*). In addition, the metabolism of carbophenothion by these species and the esterase inhibition characteristics of carbophenothion and its metabolites were studied (Machin *et al.* 1978; Martin & Steed 1978). Stanley *et al.* (1978*b*) demonstrated that the *Anser* species tested were more susceptible to poisoning by an acute dose of carbophenothion than the *Branta* geese. Studies on the level of carbophenothion in blood and the serum cholinesterase activity following a single oral dose of carbophenothion revealed that higher blood residue levels and greater inhibition of serum cholinesterase activity occurred in the *Anser* species than in the *Branta* species. An equivalent fraction of the lethal dose of carbophenothion given to the chicken and pigeon produced even larger differences with lower blood residues and less inhibition of serum cholinesterase activity compared with the *Anser* species. Thus the limited range of avian species studied in the laboratory exhibited a wide variation in their response to carbophenothion.

Elucidation of the mechanism behind this variation in response to carbophenothion in vertebrates is difficult due to the complex metabolism of the compound. All species tested appear able to metabolize carbophenothion and produce a number of products that may be more toxic than the parent compound (F.A.O./ W.H.O. 1973). Machin *et al.* (1978) have, however, concluded that the variation in the toxicity of the compound could not be explained by variation in its metabolism. It has been suggested (Machin *et al.* 1976) that the high toxicity of diazinon (*OO*-diethyl *O*-2-isopropyl-6-methylpyrimidin-4-yl phosphorothioate) to birds in comparison to mammals is related to the stability of the toxic metabolite diazoxon in avian blood. Studies have now shown that carbophenoxon, the oxon metabolite of carbophenothion, does not show any variation in stability in the blood of species which were susceptible and tolerant to carbophenothion poisoning (Machin *et al.* 1978). Lee & Pickering (1967) and Pickering & Malone (1967) demonstrated that haloxon(bis(chloroethyl)-3-chloro-4-methylcoumarin-7-yl phosphate) is very toxic to geese and suggested that this is due to the formation of a stable di-(2-chloroethyl)-phosphoryl esterase derivative and that the formation of such a derivative results in the slower reactivation of phosphorylated brain cholinesterase in geese than in other species. Martin & Steed (1978) investigated the inhibition of goose cholinesterase by carbophenoxon and demonstrated that although the cholinesterase of both *Anser* and *Branta* species differs from the cholinesterase of the pigeon and chicken there is no evidence of a slower reactivation rate for the *Anser* species.

The mechanism underlying the susceptibility of the *Anser* geese to carbophenothion poisoning in the field and laboratory has therefore not been conclusively established, although there are clearly demonstrable differences in the rate and extent of blood residue build-up at near toxic levels in closely related species of geese. Further work is in progress on the effect of carbophenothion on birds to identify the reasons for the large variation in response to carbophenothion observed between species.

Conclusions

The environmental problems experienced with the persistent organochlorine insecticide dieldrin and subsequently with the less persistent organophosphate alternatives, chlorfenvinphos and carbophenothion, have a number of implications for the registration of new chemicals. Agricultural chemicals have until recently been the only group of novel organic compounds that have been routinely examined before use for potential environmental hazard although the possible hazards from industrial chemicals are now being recognized. Chlorfenvinphos and carbophenothion were extensively tested both in the laboratory and field before clearance for use in agriculture, but the tests did not indicate that problems might arise. It is doubtful whether at that time, tests could have been designed to reveal the hazard. Pre-clearance testing is of great value in assessing the possible hazard presented by an agricultural chemical to the user of the chemical, to the consumer of the treated crop and to the environment, but such testing is always based on previous experience and is retrospective in design. Testing protocols generally can only identify features of a chemical that have been proved by past experience to be dangerous. For example, persistence is now recognized as a dangerous feature only because of the problems first encountered with the organochlorine insecticides. Recent experience with chemicals as diverse in structure and use as carbophenothion, thalidomide and vinyl chloride demonstrates that bizarre toxic effects are *a priori* difficult to demonstrate and that the retrospective nature of pre-clearance testing inevitably results in such testing occasionally being fallible. The wide variation in response by different species to chemicals and the variation in response by the same species in different physiological states necessitates that the assessment of the potential hazard to wildlife from a chemical is based both on pre-clearance testing and surveillance during the early commercial use of the chemical. The full implications of the introduction of a chemical into wide scale use in agriculture or industry cannot be completely evaluated until the chemical comes into unsupervized use.

It is essential that the fallibility of testing is recognized and that facilities are available for the thorough investigation of any problems that arise during the early use of a chemical. Such investigation demands a multidisciplinary approach to ensure that all the contributory causes are fully established, and the results must be fed back to the registration authority for immediate remedial action and subsequently into the testing procedure for other similar new compounds.

The authors are indebted to the field and laboratory staff of the Pest Infestation Control Laboratory, M.A.F.F., and of the Agricultural Scientific Services, Department of Agriculture and Fisheries for Scotland, who have participated in the investigation of wildlife problems. We also wish to acknowledge the valuable collaboration of Mr A. Machin, Central Veterinary Laboratory, M.A.F.F., and The Wildfowl Trust in the laboratory research programme.

REFERENCES (Stanley & Bunyan)

Ambrose, A. M., Larson, P. S., Borzelleca, J. F. & Hennigar, G. R. 1970 Toxicological studies on diethyl-1-(2,4-dichlorophenyl)-2-chlorovinyl phosphate. *Toxic. appl. Pharmac.* **17**, 323–336.

Atkins, T. D. & Linder, R. L. 1967 Effects of dieldrin on reproduction of penned hen pheasants. *J. Wildl. Mgmt* **31**, 746–753.

Bailey, S., Bunyan, P. J., Hamilton, G. A., Jennings, D. M. & Stanley, P. I. 1972 Accidental poisoning of wild geese in Perthshire, November 1971. *Wildfowl* 23, 88–91.

Boyd, H. & Ogilvie, M. A. 1969 Changes in the British-wintering population of the pink-footed goose from 1950 to 1975. *Wildfowl* 20, 33–46.

Boyd, H. & Ogilvie, M. A. 1972 Icelandic greylag geese wintering in Britain in 1960–1971. *Wildfowl* 23, 64–82.

Buck, W. B., Osweiler, G. D. & Van Gelder, G. A. 1973 *Clinical and diagnostic veterinary toxicology.* Dubuque, Iowa: Kendall/Hunt Publishing Co.

Bunyan, P. J. 1973 An approach to the detection of pesticide poisoning in wildlife. *Proc. Soc. analyt. Chem.* 10, 34–36.

Bunyan, P. J., Jennings, D. M. & Jones, F. J. S. 1971 Organophosphorus poisoning: a comparative study of the toxicity of chlorfenvinphos (2-chloro-1-(2′,4′-dichlorophenyl)-vinyl diethyl phosphate) to the pigeon, the pheasant and the Japanese quail. *Pestic. Sci.* 2, 148–151.

Cook, J. W. 1964 *Review of the persistent organochlorine pesticides.* London: H.M.S.O.

Cramp, S., Conder, P. J. & Ash, J. S. 1963 *Deaths of birds and mammals from toxic chemicals, September 1961–August 1962.* The third report of the joint Committee of the B.T.O. and the R.S.P.B. on Toxic Chemicals in collaboration with the Game Research Association. London: Royal Society for the Protection of Birds.

DeWitt, J. B. 1956 Chronic toxicity to quail and pheasants of some chlorinated insecticides. *J. agric. Fd Chem.* 4, 863–866.

Edson, E. F., Sanderson, D. M. & Noakes, D. N. 1963 Acute toxicity data for pesticides. *Wld Rev. Pest Control* 2, 26–28.

F.A.O./W.H.O. 1967 *Evaluation of some pesticide residues in food.* FAO: PL/CP/15; WHO/Food Add./67.32.

F.A.O./W.H.O. 1968 *1967 evaluation of some pesticide residues in food.* FAO/PL: 1967/M/11/1; WHO/Food Add./68.30.

F.A.O./W.H.O. 1972 *1971 evaluation of some pesticide residues in food.* AGP: 1971/M/9/1; WHO Pesticide Residue Series, no. 1.

F.A.O./W.H.O. 1973 *1972 evaluation of some pesticide residues in food.* AGP: 1972/M/9/1; WHO Pesticide Residue Series, no. 2.

Further review of certain persistent organochlorine pesticides used in Great Britain 1969 London: H.M.S.O.

Gaines, T. B. 1960 The acute toxicity of pesticides to rats. *Toxic. appl. Pharmac.* 2, 88–99.

Gaines, T. B. 1969 Acute toxicity of pesticides. *Toxic. appl. Pharmac.* 14, 515–534.

Graham, C. W. 1977 Personal communication.

Hagan, E. C., Jenner, P. M. & Fitzhugh, O. C. 1961 Acute oral toxicity and potentiation studies with anticholinesterase compounds. *Fedn Proc. Fedn Am. Socs exp. Biol.* 20, 432.

Hamilton, G. A. 1976 Personal communication.

Hamilton, G. A. 1977 Personal communication.

Hamilton, G. A., Hunter, K., Ritchie, A. S., Ruthven, A. D., Brown, P. M. & Stanley, P. I. 1976 Poisoning of wild geese by carbophenothion treated winter wheat. *Pestic. Sci.* 7, 175–183.

Hamilton, G. A. & Stanley, P. I. 1975 Further cases of poisoning of wild geese by an organophosphorus winter wheat seed treatment. *Wildfowl* 26, 49–54.

Hayes, W. J. 1971 *Clinical handbook on economic poisons.* U.S. E.P.A. Pesticides Programs, Public Health Service Publication 476.

Health and Safety Commission 1977 *Proposed scheme for the notification of the toxic properties of substances.* London: H.M.S.O.

Heath, D. F. & Vandekar, M. 1964 Toxicity and metabolism of dieldrin in rats. *Br. J. ind. Med.* 21, 269–279.

Hutson, D. H. & Hathway, D. E. 1967 Toxic effects of chlorfenvinphos in dogs and rats. *Biochem. Pharmacol.* 16, 949–962.

Jennings, D. M., Bunyan, P. J., Brown, P. M., Stanley, P. I. & Jones, F. J. S. 1975 Organophosphorus poisoning: a comparative study of the toxicity of carbophenothion to the Canada goose, the pigeon and the Japanese quail. *Pestic. Sci.* 6, 245–257.

Jolly, D. W. 1954 Studies in the acute toxicity of dieldrin to sheep. *Vet. Rec.* **66**, 444–447.

Lee, R. M. & Pickering, W. R. 1967 The toxicity of haloxon to geese, ducks and hens and its relationship to the stability of the di-(2-chloroethyl)-phosphorylated cholinesterase derivatives. *Biochem. Pharmac.* **16**, 941–948.

Lu, F. C., Jessup, D. C. & Lavallée, A. 1965 Toxicity of pesticides in young versus adult rats. *Fd Cosmet. Toxic.* **3**, 591–596.

Machin, A. F., Anderson, P. H., Quick, M. P., Waddel, D. F., Skibniewska, K. A. & Howells, L. C. 1976 The metabolism of diazinon in the liver and blood of species of varying susceptibility to diazinon poisoning. *Symposium on Drug Metabolism, University of Surrey* 1976, Abstract: *Xenobiotica* **7**, 104 (1977).

Machin, A. F., Cross, A. J., Howells, L. C. & Quick, M. P. 1978 Toxicology of organophosphorus pesticides: the metabolism of carbophenothion by avian and mammalian liver preparations. In preparation.

Martin, A. D. & Steed, L. C. 1978 Toxicology of organophosphorus pesticides: characteristics of the '*in vitro*' inhibition of avian brain cholinesterases by the oxygen analogue of carbophenothion. In preparation.

Mattocks, J. G. 1971 Goose feeding and cellulose digestion. *Wildfowl* **22**, 107–113.

Ministry of Agriculture, Fisheries and Food 1973 *Wheat bulb fly. Advisory Leaflet* 177. Pinner, Middlesex: M.A.F.F. (Publications).

Ogilvie, M. A. 1975 Wildfowl censuses and counts in Britain and Ireland, 1974/75. *Wildfowl* **26**, 164–165.

Ogilvie, M. A. 1977 Personal communication.

Ogilvie, M. A. & Boyd, H. 1976 The numbers of pink-footed and greylag geese wintering in Britain: observations 1969–1975 and predictions 1976–1980. *Wildfowl* **27**, 63–75.

Pest Infestation Control 1973 *Combining the report of the Infestation Control Laboratory 1968–1970 and Pest Infestation Research 1970.* London: H.M.S.O.

Pest Infestation Control Laboratory 1975 *Pest Infestation Control Laboratory Report 1971–1973.* London: H.M.S.O.

Pest Infestation Control Laboratory 1978 *Pest Infestation Control Laboratory Report 1974–1976.* London: H.M.S.O.

Pickering, W. R. 1965 The acute toxicity of chlorfenvinphos to sheep and cattle when applied dermally. *Vet. Rec.* **77**, 1140–1144.

Pickering, W. R. & Malone, J. C. 1967 The acute toxicity of dichloroalkyl aryl phosphates in relation to chemical structure. *Biochem. Pharmac.* **16**, 1183–1194.

Sanders, H. G. 1961 *The report of the Sanders Research Study Group: toxic chemicals in agriculture and food storage.* London: H.M.S.O.

Schafer, E. W. 1972 The acute oral toxicity of 369 pesticidal, pharmaceutical and other chemicals to wild birds. *Toxic. appl. Pharmac.* **21**, 315–330.

Shaffer, B. C. & West, R. 1960 The acute and subacute toxicity of technical tetram. *Toxicol. appl. Pharmacol.* **2**, 1–13.

Shakman, R. A. 1974 Nutritional influences on the toxicity of environmental pollutants. *Archs envir. Hlth* **28**, 105–112.

Stanley, P. I., Blunden, C. A., Brown, P. M. & Bunyan, P. J. 1978a Case study in ecotoxicology: wildlife problems associated with the withdrawal of the persistent organochlorine insecticide, dieldrin from use as a wheat seed treatment and its replacement by the organophosphate insecticides, chlorfenvinphos and carbophenothion. Presented at the Nato Conference on Ecotoxicology, Guildford (1977) and in preparation for *Case studies in ecotoxicology*, University of Surrey Press.

Stanley, P. I., Brown, P. M., Martin, A. D., Tarrant, K. A., Bevan, B. J., Howells, L. C. & Machin, A. F. 1978b Toxicology of organophosphorus pesticides: residues, cholinesterase activities and symptoms in pigeons, chickens and geese poisoned with carbophenothion. In preparation.

Treon, J. F. & Cleveland, F. P. 1955 Toxicity of certain chlorinated hydrocarbon insecticides for laboratory animals with special reference to aldrin and dieldrin. *J. agric. Fd Chem.* **3**, 402–408.

Tucker, R. K. & Crabtree, D. G. 1970 *Handbook of toxicity of pesticides to wildlife.* Bureau of Sport Fisheries and Wildlife, Denver Wildlife Research Center, Resource Publication no. 84.

Turtle, E. E., Taylor, A., Wright, E. N., Thearle, R. J. P., Egan, H., Evans, W. H. & Soutar, N. M. 1963 The effects on birds of certain chlorinated insecticides used as seed dressings. *J. Sci. Fd Agric.* **14**, 567–577.

Van den Heuvel, M. J. 1975 *The United Kingdom approach to the problems of assessing potential hazards to wildlife from the use of agricultural chemicals. (European Colloquium: problems raised by the contamination of man and his environment by persistent pesticides and organohalogenated compounds.)* Luxemburg: The Commission of the European Communities.

Proc. R. Soc. Lond. B. **205**, 47–61

Printed in Great Britain

The pattern of disease in the post-infection era: national trends

By Sir Richard Doll, F.R.S.

Department of the Regius Professor of Medicine,
Radcliffe Infirmary, Woodstock Road, Oxford OX2 6HE, U.K.

Information that can be used to assess trends in the health of the population is limited to the results of irregular surveys of nutritional status and 'I.Q.', to data obtained from the notification of infectious diseases, congenital malformations, blindness and other selected defects, and to mortality rates. The last have been recorded since 1841 and provide the most detailed and useful information, although they are often difficult to interpret because of changes in the nomenclature, classification, methods of diagnosis, and efficacy of treatment of disease states. In the last 40 years, mortality rates have shown progressive reductions at all ages which have continued past the time when improvements in the prevention and treatment of infectious disease might be expected to have produced their principal benefits. Notable differences have emerged between the sexes, the rates continuing to decline in women but remaining more or less stable for a period in middle-aged men. This difference can be attributed to sex differences in life-style, so that until recently the trends in women are likely to have been the better indicators of the effect of toxic agents in the environment. The available data are inadequate to assess possible effects such as alterations in behaviour, but are of some help in regard to teratogenicity and carcinogenicity.

The effect on man of man-made chemicals in the environment has to be measured against a background of effects due to many other causes. This background is itself changing kaleidoscopically, so that it may be extremely difficult to distinguish changes due to the effects of new pollutants from those due to changes in the prevalence of other factors or in clinical and public health practice.

Sources of data

The sources of data that relate to physiological status and disease prevalence are listed in table 1. Few are of any value for our purpose. Either the data relate to unrepresentative samples, or they are subjective and imprecise, or they are grossly influenced by the social factors that affect, for example, attitudes to work and the supply of facilities. Data derived from the national Hospital In-Patient Enquiry suffer also from the defect that they fail to distinguish between events that affect different individuals and those that affect the same individual on different occasions.

We may note, however, that such general indicators of health status as the age of menarche (Tanner 1973) and the growth rate of children (Tanner *et al.* 1966) have shown steady improvement, at least until the early 1960s. Regular and detailed examination of truly representative groups of primary school children have, however, been carried out only since 1972, when the National Study of Health and Growth was first established (Rona & Altman 1977). Heights and weights for the first year of the survey were very similar to those reported 12 years previously by Tanner and his colleagues, but there is reason to believe that the earlier sample came from a population that was wealthier than average and so biased upwards.

TABLE 1. SOURCES OF DATA

General Household Survey
Sickness Benefit Certificates
hospital records
    (in-patient enquiry; nosological index)
general practitioners' records
school medical examinations
morbidity registers
    (cancer; adverse reactions to drugs; blindness)
notification
    (congenital anomalies; infectious diseases)
vital statistics
    (births and deaths)
ad hoc surveys

Only one major study has been directed specifically to determine the trend in intelligence, as measured by IQ tests, and that found an increase of about 6 % between 1932 and 1947 in 11 years old Scottish children (Maxwell 1949). Such a change in a 15 year period cannot have a genetic basis. It could have been partly demographic (Cavalli-Sforza & Bodmer 1971); but the greater part was probably due to improvement in environmental factors, including schooling and nutrition.

By far the most important evidence derives from the mortality rates recorded by the Registrars General of the United Kingdom supplemented by the records of the Regional Cancer Registries and the national Scheme for the Notification of Congenital Malformations. Any effect of environmental chemicals is most likely to be seen in two groups: newborn children who are exposed during their period of development and mature adults between 45 and 64 years of age who will have had an opportunity for exposure over many years. Effects could be expected to be seen in the first group within a few years and would be seen in the latter before they were seen in older persons, because diagnoses are generally more accurate and there is less confusing background of degenerative disease from other causes; although, even so, the effect might not be seen for 10 or 20 years. I shall, therefore, examine British trends in infant and childhood mortality and in mortality at 45–64 years of age.

FOETAL AND INFANT MORTALITY

Figure 1 shows the trends in the stillbirth and infant mortality rates from 1931 to 1975. Both have continued to decline since the control of infection by sulpha drugs and antibiotics, owing to improvements in nutrition and in maternal and medical care. Tables 2 and 3 show the recent trends for the principal groups of

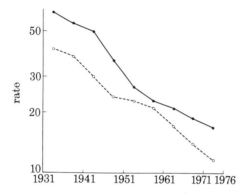

FIGURE 1. Trends in stillbirth and infant mortality rates in England and Wales: quinquennia from 1931–5 to 1971–5. ●, Infant mortality rates per 1000 live births; ○, stillbirth rates per 1000 total births.

TABLE 2. STILLBIRTH RATES (PER 1000 TOTAL BIRTHS)
BY CAUSE, ENGLAND AND WALES, 1968 TO 1973

year	diseases of mother	labour and pregnancy	congenital anomaly	other diseases of foetus
1968	2.1	6.4	2.6	3.0
1969	2.0	5.5	2.5	2.8
1970	1.8	6.0	2.5	2.7
1971	1.8	5.4	2.7	2.5
1972	1.7	5.3	2.7	2.3
1973	1.6	5.3	2.5	2.3

TABLE 3. INFANT MORTALITY RATE (PER 1000 LIVE BIRTHS)
BY CAUSE, ENGLAND AND WALES, 1963 TO 1973

year	diseases of mother	labour and pregnancy	congenital anomaly	infections	other diseases of foetus	other causes
1963	0.23	3.07	4.30	3.65	9.10	0.69
1964	0.29	2.93	4.25	3.07	8.54	0.76
1965	0.25	2.80	3.97	3.03	8.06	0.81
1966	0.20	2.63	3.89	3.06	8.33	0.74
1967	0.21	2.51	3.86	3.02	8.01	0.70
1968	0.16	2.53	3.82	2.90	8.11	0.73
1969	0.20	2.28	3.64	2.87	8.28	0.77
1970	0.17	2.27	3.71	2.71	8.66	0.65
1971	0.15	2.11	3.77	2.53	8.39	0.58
1972	0.17	2.19	3.80	2.16	8.41	0.54
1973	0.16	2.00	3.77	2.07	8.80	0.51

causes. All show a decrease, except that congenital anomalies have caused a
constant stillbirth rate since 1968. Clinically, there is no suspicion of an increase
in any life-threatening condition, except perhaps for necrotizing enterocolitis,
which has occurred more frequently in immature infants in the last decade and
has been attributed to toxic material washed out of plastic feeding tubes. This
explanation for the increase is, however, not generally accepted.

Prevalence of congenital malformations

Deaths from congenital anomalies are, of course, affected by changes in treatment, and notification rates ought to be a better guide to risk than mortality.
In practice, it is impossible to ensure complete notification of all anomalies and

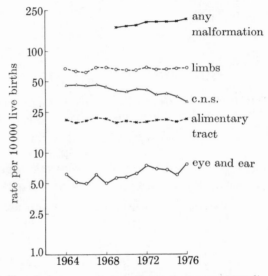

FIGURE 2. Trends in congenital malformations per 10 000 live births
in England and Wales: 1964–76.

the malformation rate that is recorded by the Registrar General is appreciably
less than that recorded by centres with a special interest in the subject. The increase in the rate of malformed babies that has been reported since 1968 (when a
new system of classification was introduced) is believed to be due to more complete reporting of minor conditions such as skin blemishes and hypospadias
(figure 2). Certainly there has been no increase in the rate of notification of any
of the major anomalies. The notification system was instituted in 1963 to provide
an early warning, should another epidemic occur like the epidemic of phocomelia
that followed the introduction of thalidomide. The thalidomide epidemic would,
however, have been missed for a long time had all limb deformities been classed
together as in figure 2, and we need to examine the trends in the notification
rates for each of a multitude of rare disorders. The Office of Population Censuses
and Surveys does this and has not observed any suspicious increase in the rate

of notification of any individual anomaly (J. A. C. Weatherall, personal communication). The reduction in the prevalence of abnormalities of the c.n.s., mostly anencephalus and spina bifida, is perhaps due to the greater use of selective abortion.

MORTALITY OF CHILDREN

At 1–14 years of age children have experienced a dramatic decrease in mortality since 1931 of nearly 90 % (figure 3). Again the decrease has continued long after the period when infections were brought under control. Nearly two-thirds of the

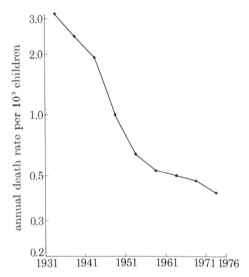

FIGURE 3. Trends in mortality in children aged 1–14 years in England and Wales, standardized for sex and age: quinquennia 1931–5 to 1971–5 (G. J. Draper, personal communication).

remaining mortality is now attributable either to violence (32 %), to neoplasms (17 %), or to congenital anomalies (15 %). With one exception, all causes of death have drifted steadily down or remained constant. The exception is asthma, which caused an epidemic of deaths in the mid-1960s. In 7 years, asthma mortality at ages 10–14 years increased seven times and in 1966 came to account for 7 % of all deaths at these ages. The epidemic is thought to have been due to the misuse of aerosol inhalers containing sympathomimetic bronchodilators and it disappeared almost as quickly as it came (Speizer *et al.* 1968*a, b*; Fraser & Doll 1971). Clinical impressions suggest that there may also have been an increase in childhood onset diabetes.

Trends in mortality from the principal types of childhood cancer are shown in figure 4 for the period 1961–75. These national data have been collected by Alice Stewart and Gerald Draper in Oxford as part of a national survey and the diagnoses confirmed by hospital enquiries (G. J. Draper, personal communication). Several types show a downward trend, due presumably to improved treatment.

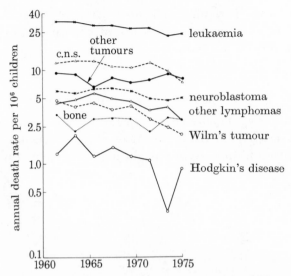

FIGURE 4. Trends in mortality from the principal types of cancer in children under 15 years of age in England and Wales, standardized for sex and age: quinquennia 1931–5 to 1951–7.

Data from single Cancer Registries are too sparse to provide reliable incidence rates, but the national data that are now being collected by Draper should enable us to observe trends in incidence, in due course, undisturbed by changes in fatality.

MORTALITY IN MIDDLE AGE

Total mortality since 1931

At 45–64 years of age the picture is very different (figure 5). Not only has the reduction in mortality since 1931 been less marked, but the extent has differed in the two sexes, the male rate being reduced by 23% and the female rate by

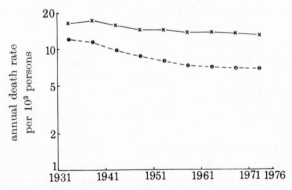

FIGURE 5. Trends in mortality in men (×) and women (•) aged 45–64 years in England and Wales, standardized for age: quinquennia 1931–5 to 1971–5.

42 %. Moreover, several causes of death, or rather diagnoses of the causes of death, have become more common. The two outstanding examples are lung cancer and ischaemic heart disease, with which we should perhaps combine the once fashionable diagnosis of myocardial degeneration. The reasons for the increase in deaths attributed to these two diagnostic groups have been discussed often and I assume that the increase is in part nosological and in part real due to changes in

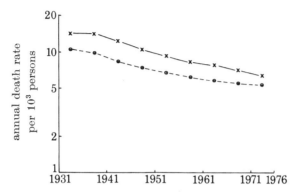

FIGURE 6. Trends in mortality from diseases other than lung cancer, ischaemic heart disease, and myocardial degeneration and similar condition in men (×) and women (•) aged 45–64 years in England and Wales, standardized for age: quinquennia 1931–5 to 1971–5.

smoking and eating habits and to a less extent to changes in physical activity. Both causes of mortality have increased to a much greater extent in men than in women and, if they are excluded, the trends in mortality are found to have been similar in both sexes (figure 6). Indeed, the male mortality rate, though still higher than the female, has declined by just over half (54 %), while the female rate has declined by slightly less (48 %). In both sexes, moreover, the rates have continued to fall until 1975.

Disease specific mortality since 1931

Examination of the trends in mortality for specific diseases is facilitated by the Registrar General, who provides on request a microfilm of the age-specific mortality rates for 1505 sex-specific causes of death, combined with the results of statistical tests for the significance of the trends. For nearly half of these causes (736) the numbers of deaths at 45–64 years of age have been too few for useful comparison.† Of the remainder, 21 have shown a significant downward trend since 1970 in at least two of the 5 year age groups (table 4), while 13 have shown a significant upward trend (table 5).

Reasons for the decrease in mortality are, for the most part, obvious. They include diagnostic artefacts (unspecified neoplasms of the c.n.s.), improvements in treatment (hypertension, cerebral haemorrhage and infections of the kidney),

† No deaths recorded in 3 or more of the last 6 years in two or more age groups.

TABLE 4. CAUSES OF DEATH BECOMING LESS COMMON IN ENGLAND
AND WALES BETWEEN 1970 AND 1975: AGES 45–64 YEARS

	sex	
cause of death	male	female
cancer of stomach	**	*
cancer of lung	*	–
unspecified neoplasm of c.n.s.	*	–
diseases of mitral valve	**	**
malignant hypertension	**	**
hypertensive renal disease	*	–
other myocardial insufficiency	**	–
cerebral haemorrhage	**	**
cerebral thrombosis	**	**
acute bronchitis	*	–
chronic bronchitis	**	**
bronchiectasis	*	–
infections of kidney	**	*
other diseases of bladder	*	–
traffic accidents (two motors)	*	–
traffic accidents (motor and pedestrian)	*	–
accidental poisoning by sedatives	*	–
accidental poisoning by gas	**	*
accident involving falling object	*	–
suicide by solid or liquid	–	*
suicide by gas	**	**

*, Significant trend ($P < 0.05$) in two 5 year age groups; **, highly significant trend ($p < 0.01$) in two 5 year age groups or significant trend in three 5 year age groups; –, no significant decrease.

TABLE 5. CAUSES OF DEATH BECOMING MORE COMMON IN ENGLAND
AND WALES BETWEEN 1970 AND 1975: AGES 45–64 YEARS

	sex	
cause of death	male	female
meningococcal infection	–	*
cancer of lung	–	**
cancer of unspecified site	–	*
non-Hodgkin's lymphoma	*	–
alcoholism	**	**
late effects of intracranial abscess	–	**
chronic ischaemic heart disease	**	**
cardiomyopathy	**	–
acute but ill defined cerebrovascular accident	**	–
peripheral vascular disease not attributed to arteriosclerosis	–	**
cirrhosis of liver	**	**
chronic nephritis	–	*
motor accidents, unspecified	–	**

*, Significant trend ($p < 0.05$) in two 5 year age groups; **, highly significant trend ($p < 0.01$) in two 5 year age groups or significant trend in three 5 year age groups; –, no significant increase.

changes in cigarette smoking (cancer of the lung), reduction in atmospheric pollution by coal smoke (chronic bronchitis), the switch from coal gas to North Sea gas (accidental poisoning and suicide by gas), and changes in prescribing habits (accidental poisoning and suicide by sedatives). Finally, several diseases have been affected by a variety of factors, some known and some unknown, which are described under the general head of an improved social condition (chronic bronchitis, diseases of the mitral valve, and cancer of the stomach). The most notable change is the decline in mortality from cancer of the stomach. The fatality rate remains high and the decline, which has occurred in many countries, particularly in North America, cannot be attributed to improved therapy. If any food contaminant or additive has been harmful, its effect has been more than counterbalanced by other factors, although it is far from clear what these may be: better preservation of food, perhaps, or increased variety of supply.

Reasons for the increase in mortality from alcoholism and cirrhosis of the liver in both sexes are obvious. The increase in cardiomyopathy and in non-Hodgkin's lymphoma may be nosological, but the former could be due to alcoholism and the latter to an unknown environmental agent.

Components of mortality in 1975

As a result of these continuing changes and the control of infection that preceded them, two-thirds of the mortality in middle age has come to be due to cancer (34 %) and ischaemic heart disease (31 %). No other group of causes contributes more than 8 % (table 6). Of these two main causes, there is more reason to fear that some cases of cancer are being produced by environmental pollutants and it may, therefore, be of interest to examine the trends attributable to individual types over a longer period. By so doing, enough data are obtained to provide useful information for cancers of all the main sites. The longer the period,

TABLE 6. PATTERN OF MORTALITY IN ENGLAND AND WALES IN 1975,
AGES 45–64 YEARS: BY SEX

cause	percentage of all deaths
cancer	34
ischaemic heart disease	31
respiratory disease	8
cerebrovascular disease	8
other circulatory disease	8
accidents, suicide, violence	4
digestive disease	3
diseases of the nervous system and sense organs	1
genito-urinary disease	1
endocrine, nutritional, and metabolic disease	1
infective and parasitic disease	1
all other diseases	1
all causes	101

however, the more difficult it becomes to distinguish trends due to changes in incidence from those due to changes in treatment and artefacts due to differences in classification and diagnosis.

Cancer mortality since 1931

The trends for cancer of 26 sites in men and 28 sites in women are shown in tables 7, 8 ,and 9 for the 40 year period 1931–35 to 1971–75. For 25 sites the age-standardized mortality at 45–64 years of age has varied by less than 1 % p.a. over the last 20 yeras.

For 16 sites the rate has decreased progressively by more than 1 % p.a. (table 7). For most of these cancers the decrease can be attributed to improved treatment, more precise diagnosis (fewer liver metastases are now recorded as primary

TABLE 7. TRENDS IN CANCER MORTALITY AT 45–64 YEARS OF AGE, ENGLAND AND WALES, 1931 TO 1975: CANCERS SHOWING A CHANGE IN MORTALITY OF LESS THAN 1 % p.a. SINCE 1951

type of cancer	sex	annual death rate per million†					change (%) 1951–5 to 1971–5
		1931–5	1941–5	1951–5	1961–5	1971–5	
cancer of							
pharynx	M	43	25	23	19	26	+13
small intestine	M	11	11	8.9	9.1	7.4	−7
small intestine	F	7.4	7.3	6.7	6.8	7.0	+4
large intestine	M	296	308	222	192	204	−8
large intestine	F	295	318	247	216	216	−13
nasal sinuses, etc.	M	—	—	11	8.3	9.5	−14
nasal sinuses, etc.	F	—	—	5.6	5.4	5.3	−5
lung	M	280	691	1497	1784	1651	+10
cervix uteri	F	—	—	213	190	177	−17
corpus uteri	F	—	—	88	73	77	−12
ovaries and Fallopian tubes	F	170	204	243	254	264	+9
vulva and vagina	F	31	26	18	17	17	−6
prostate	M	96	103	83	77	78	−6
testes	M	9.6	9.6	9.0	8.0	8.6	−4
bladder	M	92	108	125	121	115	−8
bladder	F	29	36	34	33	35	+3
kidneys and suprarenals	M	44	48	65	69	73	+12
kidneys and suprarenals	F	27	24	27	29	31	+15
brain and c.n.s.	M	87	102	146	144	140	−5
brain and c.n.s.	F	69	69	87	92	97	+11
thyroid	M	8.3	7.5	8.4	7.4	7.4	−12
Hodgkin's disease	M	29	30	35	35	31	−11
Hodgkin's disease	F	14	15	16	17	14	−12
leukaemia	M	32	44	70	77	78	+11
leukaemia	F	29	36	53	59	52	−2

† Standardized for age.

TABLE 8. TRENDS IN CANCER MORTALITY AT 45–64 YEARS OF AGE, ENGLAND AND WALES, 1931 TO 1975: CANCERS SHOWING A DECREASE IN MORTALITY OF MORE THAN 1 % p.a. SINCE 1951

type of cancer	sex	annual death rate per million†					change (%) 1951–5 to 1971–5‡
		1931–5	1941–5	1951–5	1961–5	1971–5	
cancer of							
lip	M	17	7.8	3.1	2.0	1.6	−48
lip	F	0.92	0.64	0.23	0.44	0.11	−52
tongue	M	114	45	18	11	8.6	−52
tongue	F	11	9.6	6.5	5.4	5.1	−21
mouth and tonsil	M	63	26	19	13	9.0	−53
mouth and tonsil	F	7.3	5.0	6.9	6.0	4.0	−42
stomach	M	725	685	617	503	376	−49
stomach	F	426	395	272	202	146	−47
rectum	M	298	270	194	153	152	−22
rectum	F	159	156	126	106	101	−20
liver and g.b.	M	139	102	75	61	46	−39
liver and g.b.	F	138	94	63	47	33	−48
larynx	M	108	71	53	42	36	−32
larynx	F	29	27	21	10	9.6	−54
breast	M	6.0	5.4	6.3	5.4	4.1	−35
thyroid	F	17	17	16	13	12	−25

† Standardized for age.
‡ A regular decrease of 1 % p.a. produces a decrease of 18 % in 20 years.

TABLE 9. TRENDS IN CANCER MORTALITY AT 45–64 YEARS OF AGE, ENGLAND AND WALES, 1931 TO 1975: CANCERS SHOWING AN INCREASE IN MORTALITY OF MORE THAN 1 % p.a. SINCE 1951

type of cancer	sex	annual death rate per million†					change (%) 1951–5 to 1971–5‡
		1931–5	1941–5	1951–5	1961–5	1971–5	
cancer of							
pharynx	F	12	10	13	19	17	+31
oesophagus	M	194	106	81	83	105	+30
oesophagus	F	66	51	42	44	52	+24
pancreas	M	102	109	131	147	164	+25
pancreas	F	73	69	75	84	93	+24
lung	F	69	106	168	257	397	+136
breast	F	709	651	627	687	770	+23
melanoma	M	—	—	9.4	13	20	+113
melanoma	F	—	—	11	15	26	+135
non-Hodgkin's lymphoma	M	—	—	45	55	60	+33
non-Hodgkin's lymphoma	F	—	—	25	33	41	+64
myelomatosis	M	—	—	21	31	39	+86
myelomatosis	F	—	—	16	23	26	+63

† Standardized for age.
‡ A regular increase of 1 % p.a. produces an increase of 22 % in 20 years.

tumours of the liver), or known changes in personal behaviour. Only the decrease in cancer of the stomach is wholly unexplained.

For 13 sites the rate has increased progressively by more than 1 % p.a. (table 8). The increases in cancer of the pharynx and lung in women and of the oesophagus in both sexes can be attributed to the increased consumption of alcohol and cigarettes. The explanation for the rising mortalities from cancer of the breast and melanoma is uncertain. The former is likely to be due in part to a reduction

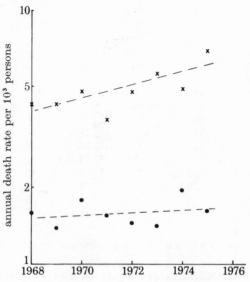

FIGURE 7. Trends in mortality attributed to pleural mesothelioma in men (×) and women (•) in England and Wales, standardized for age: 1968–1975.

in fertility, but it may also be due to an increased consumption of meat and fat. The latter has been attributed to more extensive exposure to ultraviolet light. Whether the increases in cancer of the pancreas, non-Hodgkin's lymphoma and myelomatosis are real or spurious is open to question. For myelomatosis, however, the increase has continued so long and has been so rapid that it would be unwise to ignore the possibility that part of it may be real. Apart from non-Hodgkin's lymphoma, it is the only numerically important type of cancer looking for a new environmental cause.

These data, which have been recorded over many years, are inevitably limited to cancers that were common enough in the past to be classified separately in the national records. Any cancer produced by new environmental agents that was not also produced by other causes in the past would not appear in the records as it would not have been coded separately. It is one of the problems now facing cancer registries which, if they are to be of value in monitoring the appearance of new diseases, will have to have an extremely detailed coding system capable of picking out, for example, the adenocarcinomas of the vagina in young women that were produced by giving stilboestrol to their mothers during pregnancy or the angiosarcomas of the liver attributable respectively to the medical use of thorotrast and industrial exposure to vinyl chloride. Neither of these two types of cancer can be said to have even remotely approached epidemic proportions in this country. Detailed enquiries, made retrospectively by Dr Adelstein and his colleagues in the Office of Population Censuses and Surveys (personal communication), indicate that no more than one or two cases of each have occurred annually since 1960. None of the British cases of vaginal adenocarcinoma in young women were associated with exposure to oestrogen treatment *in utero*, but one of the angiosarcomas of the liver occurred in a worker exposed to vinyl chloride, another occurred in a worker who may have been so exposed, and a third occurred in a man who resided close to a factory in which vinyl chloride was used (Baxter *et al.* 1977).†

These figures contrast with those for mesothelioma of the pleura, which may have been equally rare in the early 1950s but which now accounts for more than 200 deaths per year and, as figure 7 shows, is still becoming more common. Most cases can be attributed to occupational exposure to asbestos, and it seems likely that at least some of the others are due to asbestos particles which, for one reason or another, are now present everywhere, though only in very low concentrations, in town air.

CONCLUSION

From this and other evidence we can conclude that the health of the country has been improving steadily, and that those conditions that have become more common have for the most part done so because of personal and dietary habits that are unrelated to pollution of the environment in the normal sense of the term. It has, however, to be borne in mind that many chemicals have been introduced into the environment only in recent years and it is too early to be sure about the effects that prolonged exposure to small amounts of them may have.

Our system for recording mortality, cancer incidence, and the prevalence of congenital anomalies provides a valuable method for monitoring trends in the frequency of many of the most important pathological conditions. It cannot be expected, however, to detect the emergence of new hazards, if the lesions are specific to new agents, except by special investigations carried out restrospectively because of suspicions raised by clinicians, pathologists, and toxicologists. Other aspects of health status are much less well monitored and other methods, including the institution of regular surveys, will have to be introduced if we wish to monitor national trends in aspects such as mental health.

† The paper by Baxter *et al.* (1977) covered the period to the end of 1973. In 1974 another heavily exposed worker died of the disease.

REFERENCES (Doll)

Baxter, P. J., Anthony, P. P., MacSween, R. N. & Schever, P. J. 1977 Angiosarcoma of the liver in Great Britain, 1963–73. *Br. med. J.* ii, 919–921.

Cavalli-Sforza, L. L. & Bodmer, W. F. 1971 *The genetics of human populations*, p. 620. San Francisco: W. H. Freeman.

Fraser, P. M. & Doll, R. 1971 Geographical variations in the epidemic of asthma deaths. *Br. J. prev. soc. Med.* **25**, 34–36.

Maxwell, J. 1949 *The trend of Scottish intelligence*. London: University of London Press.

Rona, R. J. & Altman, D. G. 1977 National study of health and growth: standards of attained height, weight, and triceps skinfold in English children 5 to 11 years old. (In the press.)

Speizer, F. E., Doll, R. & Heaf, P. 1968a Observations on recent increase in mortality from asthma. *Br. med. J.* i, 335–339.

Speizer, F. E., Doll, R., Heaf, P. & Strang, L. B. 1968b Investigation into use of drugs preceding death from asthma. *Br. med. J.* i, 339–343.

Tanner, J. M. 1973 Trend towards earlier menarche in London, Oslo, Copenhagen, the Netherlands, and Hungary. *Nature, Lond.* **243**, 95–96.

Tanner, J. M., Whitehouse, R. H. & Takaishi, M. 1966 Standards from birth to maturity for height, weight, height velocity, and weight velocity: British children, 1965. Parts I and II. *Archs Dis. Child.* **41**, 454–471 and 513–635.

Discussion

R. PETO (*Radcliffe Infirmary, Oxford, U.K.*). It is commonly claimed that 70, 80 or 90 % of deaths from cancer are 'environmental' in origin, which is perhaps true in that in Britain one-third are caused by smoking, some more are caused by alcohol consumption and a large number – nobody yet knows how many – are probably caused by over-nutrition. (Giving laboratory animals as much to eat as they like has been known for 30 years to greatly increase their risk of cancer, compared with the risk among animals subjected to moderate caloric restriction.) It is also commonly believed that a large proportion of cancers are caused by environmental pollution, and this is almost certainly false. However, both in public and in private the quite proper expression of fears that certain specific environmental chemicals will prove harmful in certain specific ways is much more frequent than are reminders of the great benefits which the modern world has offered to most people in developed countries. This imbalance is so great that many are really surprised to learn that, apart from the effects of cigarette smoking, cancers and malformations are generally becoming less common, and I welcome Professor Doll's review of current trends in morbidity and mortality as a corrective to unjustified pessimism. It appears that environmental pollution is not at present a major determinant of British mortality on anything like the scale that, for example, cigarette smoking is.

R. SCHOENTAL (*Department of Pathology, Royal Veterinary College, London, U.K.*). In the tables of mortality trends from various causes in England and Wales, I found the data on gastrointestinal cancer very instructive. Mortality from stomach cancer, one of the most common types of tumour, has been steadily declining.

Not only is this good news, but it seems to me a pointer to its aetiology: due to the improvements in food storage and preservation, the levels of mycotoxins have probably declined, especially the trichothecenes, which are secondary metabolites of the common field fungi (mainly *Fusarium* species), and often contaminate cereals and other foodstuffs. In our recent experiments T-2 toxin (3α-hydroxy-4β,15-diacetoxy-8α-(3-methylbutyryloxy)-12,13-epoxy-Δ^9 trichothecene) proved to be carcinogenic to rats. It induced benign and malignant tumours of the digestive tract, of the brain, pancreas, and also cardiovascular lesions (Schoental, Joffe & Yagen, unpublished results; Schoental, R. 1977 *Cancer* **40**, 1833). 'Natural' carcinogens, such as T-2 toxin, could possibly be involved in the aetiology of diseases and tumours of the respective organs also in man, in whom such disorders occurred long before the introduction of man-made chemicals into the environment.

Proc. R. Soc. Lond. B. **205**, 63–75

Printed in Great Britain

Epidemics of non-infectious disease

By P. J. Lawther

M.R.C. Toxicology Unit, Clinical Section, St Bartholomew's Hospital Medical College, Charterhouse Square, London EC1M 6BQ, U.K.

Epidemics of non-infectious disease are often caused by exposure to industrial products, intermediates or by-products, either in the workplace or as a result of the contamination of a wider environment. Although the prime objective of research must be the recognition of the hazard and the evaluation of its magnitude so that illness may be prevented, close collaboration of clinicians, epidemiologists and toxicologists should lead to the acquisition of much knowledge of the mechanisms by which disease is caused. Catastrophes, though always regrettable, must be seen as experiments demanding careful analysis and exploitation. Many examples of different types of problem will be selected from the numerous epidemics from the time of the Schneeberg and Joachimsthal miners to the recent concern with contamination of the environs of Seveso by dioxin.

Introduction

The modest pretension of this necessarily brief contribution is that it might, by selecting for discussion a few of many epidemics of non-infectious disease, emphasize the need for collaboration of clinicians, epidemiologists, and toxicologists, in the difficult and vital tasks of identifying, assessing, forecasting and preventing long-term hazards to man from exposure to man-made chemicals in the environment. Hazards are recognized ultimately by clinicians and epidemiologists but, ideally, the toxicologist, who, by scrutiny of existing data and by further experiment, seeks to display the mechanisms by which substances exert their toxic effects, hopes as a result of his studies to be able to predict hazards and thereby avoid dangerous contamination of the environment. Clinical medicine, epidemiology and toxicology are disciplines which, practised in isolation, have obvious limitations; the need for collaboration in the study of environmental problems and the value of the results of such synergism are not so widely appreciated.

The clinical scientist is handicapped by the need, for ethical reasons, to limit his experiments on human volunteers to the study of mild, transient, easily reversible effects of exposure to suspect chemicals; moreover, his choice of subjects must be limited to healthy adults while populations at risk from contamination of the environment include the young, the old and the sick. The toxicologist who studies the effects of pollutants on animals suffers constraints which are not always recognized as being severe enough to diminish the relevance of his work to humans; the difference in reactions of various species to noxious stimuli is well recognized (though often ignored by those who seek to apply the results of the work of toxicologists) but the limits imposed by differences in life span and anatomy are less commonly recognized; the large increases in dose or concentration above those seen or expected in communal exposures, needed to produce effects in the animals' life time, may diminish the relevance of the findings; failure to recognize the importance of anatomical differences, especially of the respiratory tract in experiments on the effects of inhaled pollutants, have rendered invalid the application to man of many excellent experiments on animals. These observations or reminders are platitudinous in the context of this symposium, but when they go unheeded by those who set criteria and standards the economic consequences may be enormous.

Throughout the course of history there have been epidemics of disease due to contamination of the environment in which men work or where populations dwell; pollution may be from natural sources such as the hydrogen sulphide in Rotorua in New Zealand; man-made pollution may be accidental and catastrophic or it may be deliberate in ignorance of the harm it does or as a result of *laissez-faire*. The consequences of the contamination may be viewed merely as a matter for regret and a warning to reform; an additional, more positive, attitude is to regard such episodes as experiments that no one would be allowed to do and that offer unique opportunities for collaborative study by clinicians, epidemiologists and toxicologists. Retrospective scrutiny of past episodes must yield valuable lessons; perusal of Donald Hunter's classic work on *The diseases of occupation* and of the invaluable collections of papers in three issues of the *British Medical Bulletin* entitled 'Mechanisms of toxicity' (1969), 'Epidemiology of non-communicable disease' (1971), and 'Chemicals in food and environment' (1975), would convince anyone of the richness of this field of study. Industrial processes may cause disease by polluting the work-place and its environs; one is tempted to speculate that early epidemics may have been of silicosis in neolithic flint knappers and of farmers' lung among his agricultural contemporaries. Because of the clinical similarity of these two diseases and the commoner pulmonary tuberculosis and chronic bronchitis and emphysema, there must have been a delay in the course of time in the identification of the specific nature of these two diseases of occupation. The contaminated work-place could be seen as a most valuable laboratory in which experiments on humans have unwittingly taken place. The escape of prime products, such as lead, from industrial processes may contaminate the local surroundings and cause disease; similarly, the dissemination, accidental or deliberate, of by-products or waste material, such as mercury or cadmium, has caused infamous outbreaks of disease. Some waste products have been discharged in the genuine belief that they would cause no harm; later it has been found that harm has been caused by impurities in the waste products. The ash which collects in the heat exchangers in marine turbines powered by residual fuel oil seemed innocuous enough until it was found that workmen who were employed to remove it developed pneumonitis and other ills from the vanadium pentoxide which is a natural constituent of the pitch-like fuel used. There was no reason to believe, at one time, that the tailings from crocidolite workings could cause mesothelioma in the general population living near spoil banks. Poisoning of scientists in laboratories has produced invaluable evidence of hitherto unsuspected hazards; probably the smallest, yet not least relevant, epidemic was that which was reported by Edwards (1865, 1866) of the illness and deaths of two laboratory technicians who were working with dimethyl mercury (Frankland & Duppa 1863) at St Bartholomew's Hospital. The toxicity of mercury and its inorganic salts had long been known but the hazard posed by its organic compounds was not then recognized. The tragic story of the late recognition of the dangers of ionizing radiation by the injuries and deaths among the pioneers of radiology is well known and salutary. The side effects of therapeutic agents have sometimes been disastrous but may be regarded positively to have been experiments from which much can be learned. Almost incredibly, after the reports of the intense toxicity of dimethyl mercury mentioned above, diethyl mercury was used in 1887 for the treatment of syphilis with results which could have been forecast from Edward's observations. In much more recent times the inclusion, for no good reason, of calomel in cathartic and anti-helminthic preparations and in teething powders given to fractious infants, led to epidemics of acrodynia or pink disease which would seem to have been an idiosyncratic response to the administration of a mercury salt. The consequences of the prolonged administration of arsenic in Fowler's solution for many chronic neurological disorders produced effects which many physicians have seen. Benzene has been used for the treatment of leukaemia. The sad effects of the ionizing radiation from the diagnostic use of thorotrast and the use of radiotheraphy in ankylosing spondylitis are well known and documented. The effects on the foetus of maternal treatment with thalidamide are well known. The follow-up of the victims of Hiroshima and Nagasaki has been a cardinal example of the beneficial examination of two calamitous episodes. Natural hazards can be used for the study of the potential toxicity of environmental contaminants. The effects of natural radiation have been investigated and the results applied to the effects of exposure to man-made radiation. Lead occurs in nature and the levels of lead in the blood of populations wholly unexposed to air-borne lead derived from petrol engines or from the products of industry have been used to compare with those of urban man. Mention has been made of Rotorua; in this small town in the North island of New Zealand the atmosphere is contaminated by hydrogen sulphide from geothermal sources, and deaths occasionally occur from acute poisoning in enclosed poorly ventilated spaces where the gas accumulates (lavatories attached to petrol stations are dangerous) but there are no signs of chronic ill effects of concentrations of the gas which many environmentalists would deem to be horrifying. It may be noted with regret that there is a reluctance to publish such negative findings which might provide welcome reassurance in days when almost every factor in the environment is said to be fraught with danger.

From these general comments it should be apparent that there is a wealth of material, the study of which will enable us to understand more fully real and suspected long-term hazards due to the presence of man-made chemicals in the environment. The examples which follow are selected from a vast field and no pretence is made that within the strict limits of this informal communication it is possible to do more than illustrate the general statements made above.

CARBON TETRACHLORIDE

One might hope to be allowed, in an informal presentation to a meeting for discussion, to speak from personal experience; some of the ideas presented here were conceived while recalling working in industry in wartime under appalling conditions in badly designed plant in which the need for high production took precedence over any consideration of industrial hygiene. A stage in the manufacture of a chemical vital to defence was the chlorination of benzanilide in solution in carbon tetrachloride. Spillages and leaks were frequent and the inadequate ventilation of the plant (the inadequacy was due to the need to observe the black-out regulations) allowed the accumulation of amounts of carbon tetrachloride much in excess of the 'maximum allowable concentration'; as a result of this most of the workers suffered from 'chronic carbon tetrachloride poisoning', a feature of which was severe nausea accompanied by lassitude and other demoralizing symptoms. Turnover of labour was high and efficiency low. The problem was serious enough to merit special investigation. Much was already known of the acute toxicity of this common and useful solvent; impairment of renal and hepatic function were known to be prominent features. Stewart & Witts (1944) reported the findings of the team which investigated the problem and found no evidence of impairment of renal or hepatic function among the workers but they did see signs of hypermotility and excess irritability of the gut and concluded that carbon tetrachloride in the concentrations breathed in the works probably caused the symptoms and signs by acting on the central nervous system. Other processes, including the manufacture of magenta and dinitrophenol, were carried out in other parts of the factory; process workers could be identified by their colours: magenta and bright yellow. All of those who worked on the process involving the chlorination of benzanilide had chloracne which was usually confined to the malar area of the face, though in some cases it was much more widespread.

POLYCYCLIC AROMATIC HYDROCARBONS

It could be said that the study of occupational cancer started when Percivall Pott wrote in 1775 of cancer of the scrotum in chimney sweeps and suggested that soot might be a cause of the tumours; Butlin (1892) reported his observations that exposure to mineral oils, pitch and tar was associated with similar tumours.

Not surprisingly there was doubt about the identity of the carcinogen; the presence of arsenic, known to be carcinogenic, in coal tar was noted and blamed for the cancers by Pye-Smith (1913) and later by Bayet & Slosse (1919) who concluded, 'Le cancer arsenical et le cancer du goudron sont identiques'. In 1915, Yamagiwa & Itchikawa produced cancer on rabbits' ears by painting them with extracts of coal tar, and thereafter Kennaway (1924, 1925) and Kennaway & Hieger (1930) isolated and identified the first known carcinogenic hydrocarbons, 1,2,5,6-dibenzanthracene and 3,4-benzpyrene, from pitch. There followed many reports that left no reasonable doubt that many skin cancers in industry were due to contamination of the body with tars and oils containing these or similar hydrocarbons. When causes were sought for the alarming increase in lung cancer among the general population, attention was drawn to the excess of the disease in towns; the air of most towns in western Europe was polluted by coal smoke, and Waller (1952) demonstrated the presence of 3,4-benzpyrene in town air. Since then the overwhelming cause of carcinoma of the bronchus has been shown to be the smoking of tobacco, especially in the form of cigarettes, but there remains an urban excess which some believe to be due to the presence of polycyclic hydrocarbons in town air. An appraisal of the role of these compounds in the aetiology of lung cancer was made possible by a study of the mortality of workers in the coal gas industry (Doll *et al.* 1965) which was supplemented by a survey of contamination of the air of retort houses by polycyclic aromatic hydrocarbons (Lawther *et al.* 1965). An excess of lung cancer (less than twofold) was seen among gas workers but this was far from being proportionate to the vast excess of 3,4-benzpyrene found in retort houses when compared with that determined in town air in the days when pollution by coal smoke was high. It would seem that too much importance has been ascribed to this class of compounds as causative factors of lung cancer in the general population.

The exhaust products of the diesel engine have been blamed for the rise in deaths from carcinoma of the bronchus. Kotin *et al.* (1955) demonstrated the presence of polycyclic hydrocarbons in soot from a maladjusted diesel engine; this finding was not surprising since these compounds may be found wherever carbon-containing fuels are burned inadequately. Subsequent experiments in which animals were exposed to exhaust from diesel engines were a failure because the animals died early from carbon monoxide poisoning. There was some irony in this failure since a valuable feature of the well adjusted diesel engine is that unlike the petrol engine it produces virtually no carbon monoxide. Again there seemed to be much to be gained by going to industry; surveys of pollution in London Transport diesel bus garages were made (Commins *et al.* 1957) from which it was seen that the contribution of polycyclic hydrocarbons by diesel buses was small in comparison with that made by the coal fire and this was in accord with the findings of Raffle (1957) that there was no excess of lung cancer among workers in these garages. These figures are under annual scrutiny lest there be an effect of exhaust products which has not been manifested hitherto. The study of these carcinogenic compounds shows the value of collaboration between clinician, epidemiologist, chemist and experimental pathologist.

Asbestos

Asbestos is the name given to a group of fibrous silicates which are of great commercial value. Chrysotile, a fibrous form of serpentine which is a hydrated magnesium silicate, is also known as white asbestos and composes more than 80 % of the World's output. Among the other varieties are the various types of amphibole silicates; these include crocidolite (blue), amosite (brown) and antho-phyllite (white); there are many more. Although asbestos minerals have been known and used since ancient times, their use on a grand industrial scale dates from the discovery in Quebec and in Russia of large deposits of chrysotile about 100 years ago. There would seem to have been no good reasons to suspect that such chemically inert minerals would be harmful to man; the first case of what we now call asbestosis was observed in 1900 and later described by Murray (1907) but it was not until the late 1920s that there was enough evidence to establish an unequivocal relation between work with asbestos and the development of diffuse pulmonary fibrosis. This grim discovery prompted the enquiry which led to the classic report of Mereweather & Price (1930). Thereafter, many cases of asbestosis were reported from many parts of the industrial world. The Asbestos Industry Regulations followed in 1931 and dust control was enforced in 'scheduled' processes. Inhalation of any of the asbestos minerals can give rise to phenomena which are of little or no clinical significance; the presence of asbestos bodies in the sputum signifies nothing more than that asbestos fibres have been inhaled; likewise the presence of fibro-fatty pleural plaques commonly seen in asbestos workers are thought to be of no clinical import; calcified pleural plaques, which are endemic in parts of Finland where the soil contains much anthophyllite, are often seen in workers who have inhaled asbestos and are held by many clinicians to be harmful only when they are widespread enough to cause restriction of movement of the lungs. In the mid-1930s came the suggestion that carcinoma of the bronchus was more common in asbestos workers and this suggestion was investigated, confirmed and quantified by Doll (1955) and has been the subject of much later work. The excess incidence of carcinoma of the bronchus in asbestos workers has been shown to be greatest in those exposed over long periods, extending back to the time when concentrations were high (Newhouse 1973), and more recently the importance of cigarette smoking as a truly synergistic factor has led some to ask whether asbestos is carcinogenic in its own right or whether it merely enhances the carcinogenicity of tobacco smoke. During the war there was a report from Germany claiming that pleural 'cancers' were more frequent in asbestos workers, but it was much later (1956) that Wagner and his colleagues

noted the occurrence of large numbers of cases of pleural and peritoneal meso-
theliomata associated with exposure, often slight, to crocidolite in the northwest
region of Cape Province (Wagner 1960). These malignant tumours have since
been shown to occur often very many years after exposure to minute amounts of
crocidolite and are not dependent on the presence or absence of asbestosis. They
are rarely seen in chrysotile miners. Clearly, there seems to be some special property of crocidolite (an iron silicate) which enables it to cause these rare and fatal
tumours. At first much attention was paid to the fact that iron was substituted
for magnesium and that the difference in pathogenicity was 'chemical'; later
differing electron densities in the various fibre crystals were blamed. More recently, differences in fibre size and other physical dimensions between crocidolite
and the other types of asbestos have been studied: some animal experiments
have shown that the tendency for chrysotile, when injected intrapleurally, to
cause mesothelioma can be enhanced by altering the size and shape of the particles
so that they come to resemble crocidolite fibres.

This interest in the physical rather than chemical nature of asbestos fibres in
relation to their pathogenicity has been justified and furthered by some recent
accounts of the geographical distribution of mesotheliomas: Baris (1975) reported
an analysis of 120 'pleural mesotheliomas and asbestos pleurisies due to environmental asbestos exposure in Turkey'. Included in his series was a group of 39
patients from the villages Karain and Urgup in which asbestos had not been
found; however, the villagers use a local 'white soil' for numerous purposes and
this geologically complex mixture contains, among many other minerals, volcanic
glass fibres with dimensions similar to those of some crocidolite fibres. The geology,
epidemiology and pathological findings in this remote region are receiving much
attention. (It is of great interest to note that chest disease is familial in this region
where the very name of the village Karain means 'pain in the chest'.) Das *et al.*
(1976) have reported and discussed five cases of mesothelioma of the pleura
occurring in rural India among people with no history of exposure to asbestos
but who were engaged in the sugar cane industry; they discuss possible aetiological
factors and it is obvious that this part of the world merits the same careful investigation as is going on in Anatolia. These matters, in which clinicians, epidemiologists, and pathologists must work together, are of much more than academic
interest: when the dangers attendant on the use of asbestos, especially crocidolite,
were fully realized, advice was given to seek and use other materials with similar
physical properties but which were not known to be hazardous. The 'latent
period' for the development of mesothelioma may be long and the causation may
be related to physical rather than chemical properties of inhaled fibres; great
caution is therefore needed before assurances of the safety of asbestos substitutes
and other man-made fibres can be given. But it must be pointed out that the
results of intrapleural injection of test animals with comparatively large doses of
suspect fibres and powders must not be taken to mean that the effects seen are

necessarily applicable to man. This whole topic may be used as a model of those problems the solution of which needs the application of many skills and much wisdom.

MERCURY

Mercury is an ancient metal and there have been many accounts of epidemics of poisoning by the metal and its inorganic salts (all occupational) since Pliny described the disease of slaves who worked in mercury mines. The symptoms of poisoning include insomnia, shyness, nervousness, dizziness and tremor. The 'non-specific' – or, more properly, common – nature of the symptoms lead one to the certain belief that many cases are missed and diagnosed erroneously as psychoneuroses. The volatility of metallic mercury is popularly underestimated and one has wondered if some odd behaviour of some laboratory workers may be due to inhalation of mercury vapour derived from spilled metal which has been ignored. The diagnosis of mercury poisoning is made easier when there is a clear history of work with the metal and if the poisoning is severe enough to cause salivation and gingivitis and signs of the nephrotic syndrome. Even so, there have been patients in whom other diagnoses have been made (and wrong treatment given) through failure to take an adequate occupational history. There is therefore reason to believe that some epidemics of mercury poisoning have gone undetected. Nevertheless, perusal of accounts of poisoning among miners, gilders, mirror makers, surgeons, hatters, thermometer makers, meter menders, and detectives is rewarding if only to remind one of the importance of clinical observation in the practice of epidemiology. Pink disease, previously mentioned, has been eradicated since the use of poisonous teething powders has been discontinued. Unlike the symptoms of poisoning by mercury and its inorganic compounds, many of which disappear after exposure ceases, the damage caused by exposure to organic mercurials tends to be more serious and is less often reversible. The sad fate of the late laboratory technicians at St Bartholomew's Hospital has already been mentioned. The toxicology of the organomercurials is complex: an admirable summary is that by Magos (1975). Some organic compounds of mercury are excellent fungicides and are used for the treatment of many seeds. When the seeds are sown, the organomercurials are broken down and are made biologically inactive in the soil. Because free movement of mercury from the roots of the plant to the leaves or grain is prevented by some biochemical mechanism, there is no accumulation of mercury in the terrestial food chain. But obviously poisoning can occur in the process of formulating the seed dressing and applying it and by the consumption of treated grain dressed for sowing. Several severe outbreaks of poisoning have occurred by all these routes. Mercury can enter the terrestial food chain if birds or animals which have eaten dressed seeds are consumed. A classic early account of the symptoms and signs of poisoning by industrial exposure to methyl mercury was given by Hunter *et al.* (1940). They include generalized ataxia, dysarthria and gross constriction of the visual fields which can proceed to blindness. Memory and intelligence are said to remain relatively unimpaired in less severe cases. Fortunately, industrial cases of poisoning are now rare as the toxicity of the compounds is well recognized. But, tragically, there have been, since Hunter's description of his cases, many more opportunities to study the symptoms, signs, and pathology of the effects of alkyl mercury. Epidemics involving large numbers of people have been caused by eating bread made from wheat and other grains which had been treated with methyl or ethyl mercury. The biggest tragedy so far recorded (Bakir *et al.* 1973) occurred in Iraq during the winter of 1971/72 when, as a result of eating bread made from dressed grain, 6000 patients were admitted to hospital and more than 500 died. There had been previous epidemics from the same cause in Iraq, Pakistan and Guatemala, and smaller episodes had been reported from other countries, but only in the 1971/72 epidemic in Iraq were there quantitative studies in which exposure was related to observed clinical effects (Bakir *et al.* 1973; Kazantzis *et al.* 1976*a*; Mufti *et al.* 1976; Shahristani *et al.* 1976). These workers severally were able to assess the dose ingested, the approximate rate of elimination, and the dose–response relation for various clinical signs and symptoms.

In the aquatic environment, inorganic mercury compounds may be methylated by the action of certain bacteria and enter the food chain via fish. There have been several epidemics of poisoning by methyl mercury among fish-eating peoples but the two major epidemics occurerd in Japan in Minimata Bay (Katsuna 1968) and in Niigata (Niigata Report 1967) and were caused by the industrial release of methyl and other mercury compounds into Minimata Bay and into the Agana River after which the mercury was absorbed by fish which were then eaten. By 1971 as many as 260 cases of poisoning by methyl mercury had been reported in Minimata and Niigata, and of these 55 were fatal. More than 700 cases of poisoning had been identified in Minimata by 1974; more than 500 had been identified in Niigata. Again, these epidemics were intensively studied and, as a result, unique assessments of dose–response relations could be calculated for methyl mercury (Swedish Expert Group 1971). In addition, it was seen that the foetus was more susceptible to methyl mercury than was the mother. Details of these findings and of observations made of other epidemics of poisoning by mercury compounds are summarized admirably in *W.H.O. Environmental Health Criteria*, volume 1 (1976), from which yet again one may learn the inestimable value of the exploitation of catastrophes by clinicians, epidemiologists and toxicologists working together.

Dioxin

The compound 2,3,7,8-tetrachlorodibenzo-*p*-dioxin (TCDD) is extremely toxic and very stable. An enormous amount of work has been done on its toxicology in animals; in addition to producing chick oedema in chickens it has legion harmful effects. It has the teratogenic, foetotoxic and porphyrogenic effects which are well documented. A sensitive test of its presence is the production of hyperkeratotic lesions when painted on the ear of a rabbit. It and related chlorinated dibenzodioxins were synthesized by Tomita *et al.* (1959); it is not used commercially but it is found as a contaminant when 2,4,5-trichlorophenol is synthesized by hydrolysis of tetrachlorobenzene at high temperatures. The compound 2:4:5-trichlorophenol (2,4,5-TCP) is used to make 2,4,5-T and 2,4-D which are commonly used as effective herbicides. Dioxin may be disseminated in minute quantities as trace impurities in these compounds. There have, however, been several catastrophic releases of dioxin as a result of plant explosions when the exothermic reaction producing 2,4,5-TCP has got out of control. The most recent episode, which has set public health officials unprecedented problems and research workers a unique field for clinical and toxicological investigation, occurred at Seveso near Milan in 1976 (Giovanardi 1977) when an area of 3–4 km^2 was contaminated by a major leak from a factory making 2,4,5-TCP. The area was severely polluted by TCDD; part of the area has been evacuated. The legion effects of dioxin on man have been noted from careful observations following contamination. There was a famous episode when three horse arenas in Missouri were contaminated by spraying the ground with waste oil which had been contaminated (Carter *et al.* 1975; Kimbrough *et al.* 1977); 57 horses died. A girl 6 years old who had played in an arena became severely ill with many signs and symptoms. She was intensively studied until she recovered completely. Other persons were affected and were studied carefully (Beale *et al.* 1977). From 1960 to 1968 mixtures of 2,4-D and 2,4,5-T which were contaminated with TCDD were sprayed as a defoliant over large areas of Vietnam; Cutting *et al.* (1970) instituted studies to see whether the inadvertent exposure of the general population had given rise to an excess of birth defects. Not surprisingly, this exercise was fraught with difficulties and the interpretation of the findings are still the subject of debate. An increase in liver tumours in Vietnam had been reported by Tung (1973). The lesions seen in the more acute cases exposed to the higher concentrations resulting from local plant failure have been varied and severe. An almost common feature has been the production of chloracne. This skin lesion can be produced by exposure to several chlorinated naphtholic compounds but dioxin seems to be particularly powerful as a cause of chloracne (May 1973). There are several cases of contamination of the environment with a relatively innocuous compound (2,4,5-TCP) in which a micro-contaminant has been the cause of widespread disease. I am reminded of my wartime experience of chloracne when manufacturing a chlorinated benzanilide derivative and the indication would seem to be a retrospective analysis of such compounds to see if their capacity to cause chloracne was due to contamination by TCDD or similar compounds. There will remain the fascinating problem for the toxicologist to solve: by what mechanism is a peripheral lesion such as chloracne produced by such a compound.

POSTSCRIPT

Indulgence is sought for the platitudinous presentation of mere selection of ill-assorted instances where the collaboration of clinician, epidemiologist, and toxicologist has been essential to the elucidation of important problems in preventive medicine. The protection of the environment must surely depend upon such liaison. But lest one be tempted to think that all problems of environmental contamination might be elucidated in the near future, one is reminded that the cause of the ancient epidemic of lung cancer among the miners of Schneeberg could not have been revealed by the most careful investigations by clinician, epidemiologist and toxicologist until Marie Curie had discovered radium. Our liaison must embrace all disciplines and we must seek and hope for some new discovery which will throw more light on the mechanisms of toxicity so that we shall be able to predict hazards and so avoid them.

REFERENCES (Lawther)

Bakir, F., Damluji, S. F., Amin-Zaki, L., Murtadha, M., Khalidi, A., al-Rawi, N. Y., Tikriti, S., Dhahir, H. I., Clarkson, T. W., Smith, J. C. & Doherty, R. A. 1973 *Science, N.Y.* **181**, 230–241.
Baris, Y. I. 1975 *Hacettepe Bull. Med. Surg.* **8**, 165–185.
Bayet, A. & Slosse, A. 1919 *Bull. Acad. med. Belg.* **29**, 607.
Beale, M. G., Shearer, W. T., Karl, M. M. & Robson, M. M. 1977 *Lancet* i, 748.
British Medical Bulletin 1969 Mechanisms of toxicity. *Br. med. Bull.* **25**, 3.
British Medical Bulletin 1971 Epidemiology of non-communicable disease. *Br. med. Bull.* **27**, 1.
British Medical Bulletin 1975 Chemicals in food and environment. *Br. med. Bull.* **31**, 3.
Butlin, H. T. 1892 *Br. med. J.* i, 1341.
Carter, C. D., Kimbrough, R. D., Liddle, J. A., Cline, R. E., Zack Jr, M. M., Barthel, W. F., Koehler, R. E. & Phillips, P. E. 1975 *Science, N.Y.* **188**, 738.
Commins, B. T., Waller, R. E. & Lawther, P. J. 1957 *Br. J. ind. Med.* **14**, 232–239.
Cutting, R. T., Phuoc, T. H., Ballo, J. M., Benenson, M. W. & Evans, C. H. 1970 *Congenital malformations hydatidiform moles and stillbirths in the Republic of Vietnam* 1960–1969.
Das, P. B., Fletcher Jr, A. G. & Deodhare, S. G. 1976 *Aust. N.Z. Jl Surg.* **46**, 218.
Doll, R. 1955 *Br. J. ind. Med.* **12**, 81.
Doll, R., Fisher, R. E. W., Gammon, E. J., Gunn, W., Hughes, G. O., Tyrer, F. H. & Wilson, W. 1965 *Br. J. ind. Med.* **22**, 1–20.
Edwards, G. N. 1865 *St Bart's Hosp. Rep.* **1**, 141.
Edwards, G. N. 1866 *St Bart's Hosp. Rep.* **2**, 211.
Frankland, E. & Duppa, B. F. 1863 *J. chem. Soc.* (N.S.) **1**, 415.
Giovanardi, A. 1977 In *Proceedings of the Expert Meeting on the Problems Raised by TCDD Pollution*. Milan, 30 September and 1 October 1976, pp. 49–50.
Hunter, D., Bomford, R. R. & Russel, D. S. 1940 *Q. Jl Med.* **9**, 193.
Hunter, D. 1975 *The diseases of occupation*, 5th edn. London: English Universities Press.
Katsuna, M. (ed.) 1968 *Minamata disease*. Japan: Kumamoto University.
Kazantzis, G., al-Mufti, A. W., al-Jawad, A., al-Shahwani, Y., Majid, M. A., Mahmoud, R. M., Soufi, M., Tawfiq, K., Ibrahim, M. A. & Debagh, H. 1976 In *World Health Organization Conference on Intoxication due to Alkyl Mercury Treated Seed*, Baghdad, 9–13 November 1974, p. 37. Geneva: World Health Organization. (Suppl. to *Bull. Wld Hlth Org.* **53**.)

Kazantzis, G., al-Mufti, A. W., Copplestone, J. F., Majid, M. A. & Mahmoud, R. M. 1976 In *World Health Organization Conference on Intoxication due to Alkyl Mercury Treated Seed*, Baghdad, 9–13 November 1974, p. 49. Geneva: World Health Organization. (Suppl. to *Bull. Wld Hlth Org.* **53.**)

Kennaway, E. L. 1924 *J. Path. Bact.* **27**, 233.

Kennaway, E. L. 1925 *Br. J. med. J.* ii, 1.

Kennaway, E. L. & Hieger, I. 1930 *Br. med. J.* ii, 1.

Kimbrough, R. D., Carter, C. D., Liddle, J. A., Cline, R. E. & Phillips, P. E. 1977 *Archs envir. Hlth* **32**, 77–86.

Kotin, P., Falk, H. L. & Thomas, M. 1955 *A.M.A. Archs ind. Hlth* **11**, 113–120.

Lawther, P. J., Commins, B. T. & Waller, R. E. 1965 *Br. J. ind. Med.* **22**, 13–20.

Magos, L. 1975 *Br. med. Bull.* **31**, 241–245.

May, G. 1973 *Br. J. ind. Med.* **30**, 276–283.

Mereweather, E. R. A. & Price, C. W. 1930 *Report on effects of asbestos dust on the lungs and dust suppression in the asbestos industry*. London: H.M.S.O.

al-Mufti, A. W., Copplestone, J. F., Kazantzis, G., Mahmoud, R. M. & Majid, M. A. 1976 In *World Health Organization Conference on Intoxication due to Alkyl Mercury Treated Seed*, Baghdad, 9–13 November 1974, p. 23. Geneva: World Health Organization. (Suppl. to *Bull. Wld Hlth Org.* **53.**)

Murray, M. 1907 *Departmental Committee on Compensation for Industrial Diseases*, Cmnd 3495, p. 14; Cmnd 3496, p. 127. London: H.M.S.O.

Newhouse, M. L. 1973 *Ann. occup. Hyg.* **16**, 97–107.

Niigata Report 1967 *Report on the cases of mercury poisoning in Niigata*. Tokyo: Ministry of Health and Welfare.

Pott, P. 1775 *Chirurgical observations relative to the cataract, the polypus of the nose, the cancer of the scrotum, the different kind of ruptures and the mortification of the toes and feet*. London: Hawes, Clarke & Collins.

Pye-Smith, R. J. 1913 *Proc. R. Soc. Med.* (Clin.) **6**, 229.

Raffle, P. A. B. 1957 *Br. J. ind. Med.* **14**, 73–80.

al-Shahristani, H., Shihab, K. & al-Haddad, I. K. 1976 In *World Health Organization Conference on Intoxication due to Alkyl Mercury Treated Seed*, Baghdad, 9–13 November 1974, p. 105. Geneva: World Health Organization. (Suppl. to *Bull. Wld Hlth Org.* **53.**)

Stewart, A. & Witts, L. J. 1944 *Br. J. ind. Med.* **1**, 11–19.

Swedish Expert Group 1971 *Nord. hyg. Tidskr.* Suppl. 4, p. 65.

Tomita, N., Ueda, S. & Narisada, N. 1959 *J. pharm. Soc. Jap.* **79**, 186–92.

Tung, T. T. 1973 *Chirurgie* **99**, 427–436.

Wagner, J. C. 1960 In *Proc. Pneumoconiosis Conf.*, Johannesburg, 9–24 February 1959 (ed. A. J. Orenstein), p. 373. London: Churchill.

Waller, R. E. 1952 *Br. J. Cancer* **6**, 8–21.

Yamagiwa, K. & Itchikawa, K. 1915 *Mitt. med. Fak. K. jap. Univ.* **15**, 295.

Discussion

W. N. ALDRIDGE (*Medical Research Council Laboratories, Woodmansterne Road, Carshalton, Surrey SM5 4EF, U.K.*). It is appropriate after Professor Lawther's paper that I mention a recent episode of poisoning. During the 1976 World Health Organization Malaria Eradication Programme in Pakistan, out of over 7000 spraymen employed, 2300 became poisoned and 5 died. The insecticide used was malathion, a normally very safe material to use. From the work of W.H.O. Collaborating Centres at Carshalton, U.K., and Atlanta, U.S.A., it is now known that the dominant cause of this poisoning episode was the presence of isomalathion. This impurity potentiates the toxicity of malathion, a small percentage increasing toxicity fiftyfold. It is interesting that the phenomenon of the potentiation of the toxicity of malathion by other organophosphorus compounds was demonstrated in 1959 by experiments in animals; I believe this is the first time it has occurred in man. During the course of our investigations it has become clear that the toxicity of pure malathion is much lower than expected (10 g/kg body mass, oral to rats). Also other impurities have been found, the trimethylthiophosphates, which have undesirable toxicological properties. For example, after a single l.d.$_{50}$ dose animals may die up to 6 days later.

If I may generalize, I support Professor Lawther's statement that we must be in a position to investigate chemical accidents when they occur. To do this effectively, immediate action by experts is required in three areas: (*a*) clinical medicine and epidemiology, (*b*) analytical chemistry and (*c*) experimental biology. Only if collaboration between these three groups of experts can take place quickly will we be able to gain maximum benefit from these unfortunate accidents.

Prcc. R. Soc. Lond. B. **205**, 77–90

Printed in Great Britain

'Our load of mutation': reappraisal of an old problem

By F. Vogel

*Institut für Anthropologie und Humangenetik der Universtät Heidelberg,
Neuenheimer Feld 328, D-69 Heidelberg 1, Germany*

H. J. Muller, in a paper in 1950 entitled 'Our load of mutation', predicted the genetic decay of the human species due to increasing mutation pressure combined with relaxation of natural selection. In the meantime, much information on spontaneous and induced mutations in humans has been accumulated, and a reappraisal of Muller's conclusions gives a much less gloomy overall picture. However, a certain increase of malformation and disease can be predicted as a result of ionizing radiation and chemical mutagens. On the other hand, genetic counselling and antenatal diagnosis of genetic anomalies may help to keep the genetic risks within tolerable limits. Research on the biological conditions for the untoward effects of mutagenic chemicals considered necessary for the wellbeing of humans may also help to reduce genetic risks. The extent and kind of the risks as well as possibilities for prevention are discussed with a few examples.

'Our load of mutation'

It is now more than 25 years since Muller published a paper with this title which was to have a profound effect on research in human genetics. Muller's most important theses may be reformulated as follows:

1. A large percentage of all human zygotes are killed or prevented from reproduction by mutations.

2. The overall mutation rate per individual, i.e. the total number of new mutations contained in both germ cells which form this individual, is between 1 mutation/10 germ cells and 1 mutation/2 germ cells.

3. Every individual is heterozygous for several genes that would kill it if homozygous but are to a certain degree deleterious even in the heterozygous state.

4. Natural selection has relaxed; hence, the number of these deleterious genes in the human population increases dangerously; it may reach a critical threshold above which the whole genetic system may break down, leading to extermination of the human species.

5. The danger becomes more acute by increasing exposure to ionizing radiation.

6. We should try to curb this dangerous development by artificially regulating the human reproduction.

Was Muller right with this pessimistic view? Since he brought forward these theses, our knowledge of human genetics has been improved beyond expectations, and some of these questions raised by him, for which he had to rely on shaky extrapolations, can now be answered more confidently. In order to do so, it is useful to subdivide mutations according to the kind and the place of the genetic alteration. The conventional subdivision according to the kind of alteration gives two groups: (1) numerical and structural chromosome aberrations; (2) gene or point mutations, in which number and visible structure of chromosomes are intact; and the mutation is located in the base sequence of the DNA.

A subdivision of mutations according to the place of the alteration gives two main groups: (1) mutations in germ cells, and (2) mutations in somatic cells.

With the progress of chromosome research, and especially since syndromes with enhanced chromosome instability such as Fanconi's anaemia, Bloom's syndrome and ataxia-teleangiectasia have become known (Schroeder & Kurth 1971), we are now learning how important somatic mutations are for the cancer problem. This, however, will not be the topic of my presentation; I shall confine myself to mutations in germ cells.

How frequent are such mutations, and what are their consequences for the health of the individual? How is the natural, 'spontaneous' mutation rate influenced by ionizing radiation and chemicals in our environment? Information is relatively precise for chromosome aberrations, but much less reliable for gene mutations.

CHROMOSOME ABERRATION

Since Down's syndrome could be explained by a trisomy of chromosome 21, and Klinefelter's and Turner's syndromes turned out to be due to one X-chromosome too many or too few, respectively, a vast amount of literature on human chromosome aberrations has accumulated, and a great number of special malformation syndromes have been found to be due to numerical and structural chromosome anomalies. One property is common to almost all of them: the carriers are too severely handicapped to have children. The mutation is wiped out in the first generation after its occurrence.

Extensive chromosome studies in newborns have yielded relatively precise information on the incidence of such events. Somewhat more than one in 200 newborn infants have a (numerical or structural) chromosome aberration; in the great majority of cases, it is caused by a mutational event in the germ cell of one of the parents or in the early zygote; in very few instances one of the parents carries a structural aberration that is balanced and leads to a normal phenotype in this parent but to an abnormality in the child (Nielsen & Sillessen 1975).

However, what we see after birth is only the tip of the iceberg. About 60% of all spontaneous abortions in the first 3 months and about 10% of abortuses in the fifth and sixth months of pregnancy show a chromosomal anomaly (Lauritsen 1977). The overall incidence of chromosome aberrations among spontaneous abortions may be at least 30%, according to a very conservative estimate. About 10–15% of all pregnancies end with miscarriages. Considering this, and adding the minority of chromosome aberrations that survive into extrauterine life, we are left with the conclusion that at least 5–6 % of all human zygotes are wiped out in every generation by a severe chromosome abnormality. Very probably, this is an underestimate, as zygote loss before implantation is not considered. It is known from experiments with mice that zygote loss in the preimplantation period is appreciable and that much of it is caused by chromosome aberrations (table 1).

TABLE 1. LOSS OF ZYGOTES CONTAINING INDUCED CHROMOSOMAL
ABERRATION IN MOUSE

developmental stage of zygote	percentage of zygotes with chromosome aberrations		time after application/h	
	X-rays†	Trenimon‡	X-rays	Trenimon
M II oocytes	91.2	76.2	13	13
two-cell stage	90.4	—	48	—
blastocysts	—	40.3	—	105
dominant lethal assay, 13.5 days p.c.	72.3	41.1	324	324
chromosome analysis of 13.5 day old embryos	1	0.38	324	324

Female mice 8–12 weeks old were treated with X-rays (200 R; 51.6 mC kg^{-1}) or Trenimon, a cytostatic agent (25 mg/kg).
† Reichert (1976). ‡ Basler *et al.* (1976).

GENE MUTATIONS

Next, gene or point mutations may be considered, which Muller had primarily in mind when he estimated our load of mutation. There is information at two levels: the mutation rates per phenotype are known for a number of rare hereditary diseases with autosomal-dominant or X-linked recessive modes of inheritance, and the molecular mechanism of a great number of mutations affecting the haemoglobin molecule has been studied. So far, however, the two levels have almost no connection with each other. Most mutations, if analysed at the molecular level, turn out to be due to a single base substitution in the transcribing DNA strand; however, molecular deletions, chain elongations and products of abnormal recombination have also been observed (Lehmann & Huntsman 1974). It should, in theory, be possible to calculate from such data mutation rates per codon; and, indeed, two very preliminary estimates coincide in the order of magnitude, which seems to be *ca.* 10^{-8} to 10^{-9} per codon per generation. Large-scale population studies with electrophoretic methods that would be needed for more precise rate determinations at this level are, however, still lacking. One group in the United States is now actively preparing such a large-scale and long-term effort. There are two main problems in such an endeavour (Neel *et al.* 1973): (1) methods have to be developed that make detection of mutation-determined protein variations possible for a sufficiently representative number of structural genes; and (2) blood sampling of hundreds of thousands of children, and additional family examinations in cases in which a protein variant could indicate a new mutation, has to be organized.

The technical and financial difficulties are formidable. However, if such a programme were to be carried through on a sufficiently large scale and for a longer time period, it would help us not only to gain valuable information on spontaneous mutation, but also to develop a monitoring system for unknown environmental mutagens. Not all mutations that can be detected by such a system would increase our mutational load, as Muller had in mind; many of them would only add to the genetic variability within our population, and would be selectively more or less neutral – at least under modern living conditions – or would even enhance the adaptibility of our species to varying environmental conditions.

When we talk about the impact of mutations on human welfare, we should look rather at the dominant and X-linked mutations that lead to clear-cut, and in most cases severe, hereditary diseases. Screening of the literature for such mutation rate estimates renders fairly reliable figures, in some cases from two or more different populations, for 13 dominant and five X-linked conditions. They range from 10^{-4} to 10^{-7} per disease per generation. About 1000 different dominant and X-linked conditions are known. Unfortunately, however, the extrapolation from known to unknown mutation rates is not feasible, as the hereditary diseases for which we know the mutation rates are not an unbiased sample of all dominant and X-linked conditions. Most mutation rates for single diseases may be much lower.

Still, from evidence of various kinds, the overall *incidence* of autosomal dominant or sex-linked hereditary diseases can be estimated as *ca.* 1 % of all newborn infants. How much of this burden is created anew in every generation by mutation cannot be estimated reliably. It could be 5 % or it could be 50 %.

Unlike autosomal dominant and X-linked recessive mutations, nothing is known about mutation rates leading to autosomal recessive diseases. An autosomal recessive mutation leads to a hereditary disease only if it happens to meet in an individual another mutation at the same gene locus. In most cases, it will be present in the heterozygous state. For some of the more common recessive diseases, the heterozygote frequency can be calculated, and may be surprisingly high. In a white U.S. American or northwestern European population, for example, about one individual in 40 or 50 is heterozygous for the phenylketonuria gene, and even one in 20 or 30 is heterozygous for cystic fibrosis of the pancreas. Now, close to 1000 such recessive diseases are known. At first glance, this should lead us to suspect that, as Muller assumed, each of us may be heterozygous for several such recessive genes. However, most of the known autosomal-recessive diseases are very much rarer than the above-mentioned examples, many of them having been observed in a few single families only, or in small and isolated populations in which their gene frequencies had been boosted by genetic drift. Not all recessive mutations, on the other hand, lead to clearly identifiable phenotypes, and hence to known recessive diseases. Some may kill the zygote before birth, acting as lethal factors.

CONSANGUINITY STUDIES

One way of estimating the number of such recessive genes, whether leading to hereditary diseases or to early zygote death, is a comparison of offspring from consanguineous and non-consanguineous marriages. First cousins, for example, have one-eighth of their genes in common by descent, and if one of these genes is a recessive lethal or leads to a recessive disease, it may become homozygous in a child. Hence, the number of homozygotes for such genes should increase with increasing degree of kinship of the parents, and theoretically, such a comparison should even enable us to estimate the average number of such recessive genes per individual.

In the past 20 years or so, human population geneticists have tried on a large scale and in various populations to study such consanguinity effects. The overall results were disappointing: no reliable estimate of the number of recessive genes per individual emerged. This is mainly due to the fact that in almost all populations, individuals who marry a close relative tend to live in socio-economic conditions that are different from those of other individuals; in most cases less favourable, but sometimes even better. However, socio-economic factors influence perinatal and infant mortality, parameters that were primarily investigated in most of these studies. Such factors should not influence the frequency of known recessive disease. However, the consanguinity studies have not even given a clear picture of how frequent such diseases may be in the progeny of consanguineous couples. Either most of these studies have not managed to ascertain these diseases and to diagnose them correctly, or recessive diseases are indeed so rare even in the progeny of consanguineous marriages that the sample sizes of most studies were too low for a determination.

One conclusion, however, seems to be justified: Muller's estimate of about eight recessive deficiency genes for which every individual is supposed to be heterozygous is much too high if we consider only those mutations which in the homozygous state leads to well known recessive diseases. For the frequency of recessive mutation leading to foetal and perinatal death, not to speak of those killing the zygote in a certain percentage of cases during infancy and childhood, nothing is known with certainty. However, the evidence, ambiguous as it may be, rather favours the view that this number may also be lower. Very probably, many such lethals can become effective only under certain environmental conditions; a concept of lethal factors derived mainly from observations on fruit flies kept in bottles is inadequate to describe the situation in humans.

I suspect that a number of recessive genes, when homozygous, will kill the zygote at a very early stage of development – either before, or a short time after, implantation in the uterus. This zygote loss has gone unnoticed in the consanguinity studies, just as it goes unnoticed in the life of the woman to whom it occurs. A short delay of menstruation, then a bleeding that might be somewhat heavier than usual – this happens very often, and is forgotten after a few weeks.

HEALTH STATUS OF HETEROZYGOTES: INFLUENCE OF THE ENVIRONMENT

It was Muller's concern that the genes for which every individual was supposed to be heterozygous would reduce wellbeing even in the heterozygous state. This concern was, again, founded on observations in *Drosophila* that suggested a reduction of, on the average, *ca.* 2–4 % of fitness in heterozygotes for recessive lethals as compared with the wild type. In the meantime, heterozygote tests have become available for many human diseases, recessive or X-linked (Hsia 1969). These heterozygotes are usually quite healthy. For most human enzymes, half the amount of normal activity is sufficient for normal function. However, there are exceptions. A heterozygote for cystic fibrosis, for example, may faint earlier in hot and sticky weather, as he loses too many electrolytes in his sweat. Or a heterozygote for myoklonus epilepsy may run a somewhat higher risk of getting an atypical kind of epilepsy. Especially well analysed are the defective variants of the enzyme alpha-1-antitrypsin: homozygotes have a high risk of suffering from obstructive emphysema of the lungs, whereas heterozygotes may also develop emphysema in middle age, if they smoke too much or get repeated bronchial infections for other reasons (Kueppers 1975).

This example shows the interplay of genetics and environment in a human mutation that may be, but need not be, a threat to human welfare.

OVERALL FREQUENCY OF HOMOZYGOTES FOR RECESSIVE CONDITIONS

We should, however, keep in mind our main question: how frequent are homozygotes of recessive genes in human populations? Here, the differences between populations are very high. In northwestern European or North American white populations, evidence from medical genetics suggest an overall incidence at birth of between about 1/1000 and 1/5000 newborn infants. In an American Black population, this figure would be higher, about 2/1000 or 3/1000, owing to sickle cell anaemia. In some isolates, such as the Amish in the U.S.A. or inhabitants of the island Mljet in Yugoslavia, even higher frequencies are observed.

At first glance, the overall frequency in a well mixed, random mating population such as the white U.S. population seems to enable us to estimate also the overall mutation rate for such recessive genes. With every patient who suffers from a recessive disease and is therefore unable to reproduce, two abnormal genes are eliminated from the population, and if an equilibrium between mutation and selection exists, about the same number of mutations is expected to arise anew by mutation. Unfortunately for the population geneticists – and fortunately for our present-day families – most modern populations are not in equilibrium for recessive mutations. This is due to a change in marriage customs. In the not too remote past, human populations used to be subdivided into subgroups with much in-group inbreeding and little between-group gene flow arising from migration and intermarriage. In these isolates, some recessive genes became frequent owing to genetic drift, whereas others were very rare or not present at all. At the same time, marriages between genetically related individuals led to relatively frequent homozygosity among the offspring, and to elimination of recessive genes by hereditary disease.

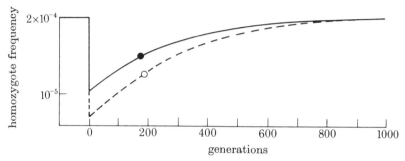

FIGURE 1. Decrease in the frequency of recessive homozygotes by the complete abandonment of inbreeding, and gradual increase to the old value as the result of an excess of mutants freshly produced over those eliminated. Mutation rate = 10^{-5}; selection coefficients for the recessive homozygotes $(s) = 0.5$; inbreeding coefficients, $F = 0.003$ (—) and $F = 0.005$ (– – –). At the points ● and ○, respectively, the gene frequency, q, reaches one-half the difference to the new equilibrium frequency.

At present, however, the boundaries of these isolates are breaking down (Jacquard 1974). This leads to a levelling-out of gene frequencies at a lower level, by which the probability of homozygosity is reduced even under random mating as homozygote frequency is the square of the gene frequency. At the same time, however, the frequency of consanguineous marriages has dropped down sharply, leading to an additional decrease of homozygosity for rare recessive mutations (figure 1).

In conclusion, then, because of the break-up of isolates and the decrease of consanguineous marriages we are enjoying at present an unusually low incidence of autosomal-recessive diseases and other ill effects of homozygosity for rare genes. This is fortunate for our society, and is, by the way, a strong argument in favour of intermarriage between races and ethnic groups. It is unfortunate for the population geneticists, as it removes every sound basis for estimation of a mutation rate for such recessive genes (Vogel & Rathenberg 1975). Mutation rate estimates have been published for single recessive genes as well as for groups of recessive mutations. They depend on assumptions that are demonstrably wrong, and hence have no scientific foundation.

MULTIFACTORIAL DISEASES

So far, mutations have been discussed that lead to visible changes in number or structure of chromosomes, or to a well defined hereditary disease. However, genetically determined ill effects can be identified in many other diseases, which we call, for want of a more precise identification, multifactorial. Most of these conditions tend to be somewhat more often concordant in monozygotic than in dizygotic twins or other first-degree relatives; and relatives, in turn, tend to be more often affected in a similar way than individuals taken at random from the population. Usually a genetic predisposition interacts with unfavourable environmental conditions.

With the exception of those of us who will die from cancer due to a somatic mutation that is caused by an environmental agent, almost everybody in this room will die some day from such a multifactorial disease, be it cardiac infarction, apoplexy due to high blood pressure, or whatever. This may demonstrate what kind of difficulty we run into as soon as we try to define this group. Yet we shall try at least a crude classification. There are three main groups: (1) congenital malformation, except for those caused by single genes or chromosome aberrations; (2) so-called constitutional diseases, for example diabetes, rheumatic diseases, hypertonia, and many heart diseases; and (3) mental deficiency and mental illness.

The incidence of malformations among newborns is usually given as a small percentage; the figures in the literature differ vastly, depending on whether small or almost trivial malformations are included. For really severe malformations, about 1–2 % may be a fair estimate for the average risk. However, differences occur between populations. A higher incidence of congenital dislocation of the hip in the Bretagne, in northwest France, may have genetic causes (see Jacquard 1974), while the high incidence of spina bifida and anencephaly in the west of Great Britain may be due to some unknown environmental factor.

In recent years, medical geneticists have managed to isolate from the big group of malformations clinico-genetic subtypes that are obviously caused by single dominant genes, in most cases with reduced penetrance and very variable expressivity. Experiences with radiation-induced skeletal malformations of mice suggest that rather ill-defined phenotypes with combinations of skeletal and other organ malformations might also occur in humans, as a result of spontaneous mutation (Selby 1977). However, nobody can tell how frequent they are, and what proportion of all malformations are on the contrary caused by environmental agents acting during embryonic development.

The situation is still worse for the so-called constitutional diseases. Their frequency shows strong variations with age and with living conditions, for example nutrition. Under the affluent living conditions of northwest Europe and the U.S.A., for example, about 10 % or more of the population above 60 years old have diabetes mellitus. Aged people frequently suffer not only from one, but from two or more such constitutional diseases.

The third group is mental deficiency or mental illness. About $\frac{1}{4}$–$\frac{1}{2}$% of all new-born infants will develop severe mental deficiency, and about 2–3% will be mentally subnormal (Penrose 1963). The risk estimates for mental illness run at about 1% for schizophrenia and 0.4% for the whole group of manic–depressive diseases and endogenous melancholias.

For all these malformations and diseases, a certain genetic component has been shown. However, the exact genetic mechanisms are largely unknown. Consequently, it is also unknown whether and to what degree incidence of these diseases is maintained in our population by spontaneous mutation. For many of these diseases, associations have been shown with genetic polymorphisms. For some types of cancer, for example, as well as for thromboembolic diseases, individuals with blood group A have a higher risk than those with group O, who are more often affected with duodenal ulceration (Vogel & Helmbold 1972). Multiple disease associations have also been described for the major histocompatibility system, especially antigens of the HLA-B locus (Dausset & Svejgaard 1977). These genetic polymorphisms, however, are not maintained by mutations in our populations but are most probably the result of natural selection in the past that has favoured either the heterozygotes (heterosis), or rare genotypes and genetic variability as such (frequency-dependent selection; for details on the modes of selection see Jacquard (1974)).

This has induced some authors to believe that all genetic variability for disposition to such common diseases may be due to various combinations of polymorphic genes, and that none of it is maintained in our population by mutation pressure; and, consequently, that no increase in mutation rate could enhance incidence of multifactorial diseases.

This is a very optimistic viewpoint. Unfortunately, however, within the big groups of malformations and constitutional diseases, rare types that are due to fairly recent mutation and may have, for example, a dominant mode of inheritance with complete penetrance only have a chance to be singled out if they happen to be combined with specific and conspicuous clinical signs. An unknown number of them may therefore be hidden within the comprehensive and ill-defined groups of common malformations, constitutional diseases, and conditions leading to mental deficiencies and mental diseases of unknown origin.

After many discussions, an expert group of the International Commission for Radiation Protection recently concluded that the mutational impact of such genes on the population might be of the order of magnitude of the impact of chromosome mutations (except those working before birth) and dominant and X-linked mutations together. Let us live with this assumption until we will have more reliable data.

THE IMPACT ON HUMAN WELFARE OF SPONTANEOUS MUTATION

In conclusion, the approximate incidence of death or severe disease arising from spontaneous mutation can be estimated (table 2). With the exception of chromosome aberrations, the additional mutational loss of zygotes during early development cannot be estimated. However, while miscarriage – and especially repeated miscarriage – might be a great misery for the afflicted family, a certain amount

TABLE 2. ESTIMATED INCIDENCE OF DEATH OR SEVERE DISEASE
DUE TO SPONTANEOUS MUTATION

type of mutation	incidence
chromosome aberrations, numerical and structural	*ca.* 0.5 % (newborn); over 5 % (including miscarriages)
dominant and X-linked diseases	less than 50 % of *ca.* 1 %: less than 0.5 %
autosomal recessive hereditary diseases and lethal genes	unknown
multifactorial diseases (malformations; constitutional diseases; mental retardation or illness)	1 %?

of spontaneous abortion due to genetic causes can hardly be considered too severe an impact on human welfare at a time when the overwhelming social problem of the world is overpopulation. Recently, it has even been suggested that the relatively high frequency of abortions in humans may have been a genetic adaptation carrying with it a selective advantage for the population as a whole, in that it may have helped to prevent women in primitive human societies from having too many successive births of healthy children that not only challenged the lives of the mothers but also endangered the infants who had to be weaned too early from breast feeding (Jacobs 1975).

INDUCED MUTATION

Consideration of the impact on human health of mutations induced by ionizing radiation or chemical mutagens must take into account this background of naturally occurring mutations. No mutagenic agent creates anything anew; all mutagens known so far enhance the probability of events that can also occur spontaneously. According to current estimates on the basis of data from mice (Lüning & Searle 1971), the present average population exposure to artificially applied ionizing radiation may lead to a mutation rate increase of about 2–5 %. This is the price which we pay for the benefit of X-ray diagnostics, to mention only one example. Such an increase does not imply a threat to our whole genetic system. On the contrary, it would be hard, if not impossible, to demonstrate it by epidemiological methods (Vogel 1970). It nevertheless means severe sickness and suffering for many human beings.

As to chemical mutagens, no data are available that would allow us to estimate even the approximate order of magnitude of a mutation rate increase due to environmental agents. The notion that all chemicals should be tested for mutagenic activity before they are introduced for use in humans is now generally accepted. However, the methods to be applied are still a matter of discussion. In my opinion, this discussion suffers from the fact that most scientists working in this field have a biological, not a medical, training. They underestimate the pecularities of mammalian germ cell development. Let me give one example: INH, the most common drug used against tuberculosis, was found to be mutagenic in the so-called host-mediated assay (Röhrborn *et al.* 1972) and in a bacterial system *in vivo* with liver microscomes (Herbold & Buselmaier 1976). This triggered a large-scale cooperative study of 13 laboratories with almost all available cytogenetic *in-vivo* methods, with a negative result (Röhrborn *et al.* 1978). Finally, studies *in vitro* on a single cell system revealed that INH is not mutagenic in itself but is a post-replication repair inhibitor (Klamerth 1978). This example shows how complex the problem may be. So far, there is unambiguous evidence for mutagenic activity mainly for cytostatic agents. However – and this is an important practical result of experimental research with mutagens in animals – much of the genetic danger from these agents can be avoided if, in both sexes, no treatment is given during the last 8 weeks or so before conception, and, in the woman, during pregnancy (Fuhrmann & Vogel 1976).

Let us have a second look at the theses of Muller.

His first thesis was that a large percentage of all human zygotes are killed or prevented from reproduction by mutations.

His second thesis was that the overall mutation rate per individual, i.e. the total number of new mutations contained in both germ cells which form this individual, is between one mutation per ten germ cells and one mutation per two germ cells.

Considering all the known chromosome mutations, adding those that go unnoticed because they kill the zygote before implantation, adding also the point mutations leading to dominant or X-linked hereditary diseases, and allowing for the unknown number of recessive mutations, the conclusion can hardly be avoided that Muller was right for these two theses.

His third thesis was that every individual is heterozygous for several genes that would be lethal if homozygous and are to a certain degree deleterious even in the heterozygous state. This thesis is clearly wrong for clear-cut recessive diseases with well defined phenotypes. Moreover, studies on reproduction and child mortality in consanguineous marriages, though inconclusive because of socio-economic biases, are more compatible with the view that the additional burden of recessive lethals with such effects as infant mortality, may be much lower than assumed by Muller. So his third thesis is most probably too pessimistic.

His fourth and fifth theses were that natural selection has relaxed; hence, the number of deleterious genes increases; a critical threshold may be reached above which the genetic system may break down. The danger becomes more acute with increased exposure to mutagenic agents.

These theses are wrong for chromosome aberrations. Preventive or therapeutic methods have so far neither helped to save zygotes that would otherwise have been aborted, nor enabled bearers of unbalanced chromosome aberrations to reproduce. On the contrary, widespread use of methods for antenatal diagnosis will soon appreciably reduce the impact on human welfare of genome and chromosome mutations.

Muller's thesis is correct, on the other hand, for some dominant and X-linked recessive diseases for which effective therapies have become available. Examples are retinoblastoma, the eye cancer of children that can now be removed in most cases, and haemophilia A, which is now being treated by substitution of the missing coagulation factor. However, the awareness in many populations of planned parenthood and the merits of genetic counselling is increasing rapidly. Artificial selection by planned reproduction will therefore soon replace natural selection.

For some of the mutations with less well defined phenotypic consequences, Muller's thesis may also be correct. For example, complex functional systems such as vision or the immune system may gradually deteriorate owing to repeated mutations that are no longer wiped out by natural selection. However, some problems should be left for the human geneticists of future generations, who may gradually identify gene mutations that are responsible for such phenotypes, and find ways for preventing their ill effects.

As to Muller's sixth thesis, we are already regulating our reproduction artificially, and will do so more efficiently in the future. What is needed is further improvement of our knowledge of genetic diseases in humans, especially those that are still ill defined genetically.

If Muller were still alive, he would be pleased that, while some of his predictions were proved correct, the overall picture for the genetic future of the human species looks much more optimistic. Our load of mutation is not negligible, especially if we consider the fate of the individual and the family which may be severely affected. However, we can do something about it, both by genetic counselling, and by avoiding unnecessary mutations.

REFERENCES (Vogel)

Basler, A., Buselmaier, B. & Röhrborn, G. 1976 Elimination of spontaneous and chemically induced chromosome aberrations in mice during early embryogenesis. *Hum. Genet.* **33**, 121–130.

Dausset, J. & Svejgaard, A. 1977 *HLA and disease.* Copenhagen and Baltimore: Munksgaard/Williams & Wilkins.

Fuhrmann, W. & Vogel, F. 1976 *Genetic counseling*, 2nd edn. New York, Heidelberg and Berlin: Springer-Verlag.

Herbold, B. & Buselmaier, W. 1976 Induction of point mutations by different chemical mechanisms in the liver microsomal assay. *Mutat. Res.* **40**, 73–84.

Hsia, D. Y.-Y. 1969 The detection of heterozygous carriers. *Med. Clins N.A.* **53**, 857.

Jacobs, P. A. 1975 The load due to chromosomal abnormalities in man. In *The role of natural selection in human populations* (ed. F. M. Salzano), pp. 337–352. Amsterdam: North-Holland.

Jacquard, A. 1974 *The genetic structure of populations.* Berlin, Heidelberg and New York: Springer-Verlag.

Klamerth, A. 1978 Inhibition of post-replication repair by isonicotinic acid hydrazide. *Mutat. Res.* **50**, 251–261.

Kueppers, F. 1975 α_1-Antitrypsin. In *Humangenetik, ein kurzes Handbuch* (ed. P. E. Becker), vol. 1, pt 3, pp. 35–49. Stuttgart: Thieme Verlag.

Lauritsen, J. G. 1977 Genetic aspects of spontaneous abortion. *Dan. med. Bull.* **24**, 169.

Lehmann, H. & Huntsman, R. G. 1974 *Man's hemoglobins.* Amsterdam and Oxford: North-Holland.

Lüning, K. G. & Searle, A. G. 1971 Estimates of the genetic risks from ionizing radiation. *Mutat. Res.* **12**, 291–304.

Muller, H. J. 1950 Our load of mutation. *Am. J. hum. Genet.* **2**, 111–176.

Neel, J. V., Tiffany, T. O. & Anderson, N. G. 1973 Approaches to monitoring human populations for mutation rates and genetic diesease. In *Chemical mutagens* (ed. A. Hollaender), vol. 3, pp. 105–150. New York: Plenum.

Nielsen, J. & Sillesen, I. 1975 Incidence of chromosome aberrations among 11 148 newborn children. *Hum. Genet.* **30**, 1–12.

Oftedal, P. *et al.* 1976 *I.C.R.P. commission.* (In the press.)

Penrose, L. S. 1963 *The biology of mental defect*, 3rd edn. London: Sidgwick & Jackson.

Reichert, W. 1976 Strahleninduzierte Chromosomenanomalien in der Oogenese der Maus und deren Verhalten in der Embryogenese. Dissertation, Heidelberg.

Röhrborn, G., Propping, P. & Buselmaier, W. 1972 Mutagenic activity of isoniazid and hydrazine in mammalian test systems. *Mutat. Res.* **16**, 180–194.

Röhrborn, G. *et al.* 1978 Correlated studies of the cytogenetic effect of isoniazid (INH) on cell systems of mammals and man. *Hum. Genet.* (In the press.)

Selby, P. B. & Selby, P. R. 1977 Gamma-ray-induced dominant mutations that cause skeletal abnormalities in mice. II. Description of proved mutations. *Mutat. Res.* **43**, 357–375.

Schroeder, T. M. & Kurth, R. 1971 Spontaneous chromosomal breakage and high incidence of leukemia in inherited disease. *Blood* **87**, 96–112.

Vogel, F. 1970 Monitoring of human populations. In *Chemical mutagenesis in mammals and man*, pp. 445–452. Berlin, Heidelberg and New York: Springer-Verlag.

Vogel, F. 1975 Approaches for an evaluation of the genetic load due to mutagenic agents in the human population. *Mutat. Res.* **29**, 263–269.

Vogel, F. & Helmbold, W. 1972 Blutgruppen-Populationsgenetik und Statistik. In *Humangenetik, ein kurzes Handbuch* (ed. P. E. Becker), vol. 1, pt 4, pp. 129–557. Stuttgart: Thieme Verlag.

Vogel, F. & Rathenberg, R. 1975 Spontaneous mutation in man. *Adv. hum. Genet.* **5**, 223–318.

Discussion

A. G. SEARLE (*M.R.C. Radiobiology Unit, Harwell, Didcot OX11 0RD, U.K.*). I should like to endorse Professor Vogel's view that if we want to assess the human genetic risks which might arise from use of a particular chemical we must use relevant data from man himself or from related experimental mammals. Information from lower organisms is useful for initial screening but cannot tell us about actual risks. More work on dominant mutations in mammals and on non-disjunction, leading to trisomy, seem of particular importance at present.

As far as information from man himself is concerned, I should like to ask

Professor Vogel which methods of monitoring human groups he thinks are most promising.

F. VOGEL. I agree in all aspects with Dr Searle's remarks. As to monitoring human groups, I should think that monitoring for chromosomal aberrations at birth would be the most feasible method.

Proc. R. Soc. Lond. B. **205**, 91–110 (1979)

Printed in Great Britain

Congenital malformations and other reproductive hazards from environmental chemicals

By F. M. Sullivan and S. M. Barlow

*Department of Pharmacology, Guy's Hospital Medical School,
St Thomas Street, London SE1 9RT, U.K.*

From a number of disasters which have already occurred throughout the world, it is known that the reproductive process in both animals and man may be severely affected by chemicals. The range of effects that might occur include not only foetal death or malformation, but also effects on the subsequent development, behaviour, intelligence and reproductive capacity of offspring which appear otherwise normal at birth. The special sensitivity of the foetus to some environmental carcinogens is also discussed. Some of the problems in screening for such effects in animals are mentioned along with the need for adequate monitoring programmes to detect reproductive toxicity both from industrial exposure to chemicals and from more general environmental exposure.

In both animals and man, the reproductive process may be severely affected by environmental chemicals. It is not only exposure during pregnancy itself that may be dangerous, but exposure of males or females to environmental chemicals at any time may cause abnormalities in the germ cells resulting in infertility or an adverse outcome for a subsequent conceptus. The range of possible hazards to the foetus (see table 1) includes not only intrauterine death, spontaneous abortion, and congenital malformation, but might also involve subtle effects on the subsequent development, behaviour, intellectual and reproductive capacities, and general health of offspring which appear otherwise normal at birth. The difficulties of recognizing such late-emerging effects, where the appearance of the

TABLE 1. RANGE OF POSSIBLE REPRODUCTIVE EFFECTS OF
ENVIRONMENTAL CHEMICALS

before conception	during pregnancy	after delivery
menstrual disorders	maternal	abnormal development from
male sexual potency and libido	enhanced toxicity	chemicals transmitted in
infertility	toxaemia	breast milk or brought home
germ cell mutation	abortion	on parent's work-clothes or
	foetal	in the environment
	death	
	malformation	
	functional deficit	
	biochemical change	
	growth retardation	
	mutation	
	cancer	

defect may well be some time removed from precipitating events, should not be underestimated. Long-term studies involving large numbers of subjects may be necessary to detect late-emerging or small effects. The view, not uncommonly heard, that if environmental chemicals were having any significant adverse effects they would have been noticed by now, is therefore clearly untenable.

TABLE 2. TOXIC SUBSTANCES CONSIDERED TO BE EXTRA HAZARDOUS DURING
PREGNANCY (U.S. DEPARTMENT OF LABOR 1942)

aniline	nitrobenzene and other nitro
benzene and toluene	compounds of benzene and
carbon disulphide	its homologues
carbon monoxide	phosphorus
chlorinated hydrocarbons	radioactive substances and
lead and its compounds	X-rays
mercury and its compounds	turpentine

Other toxic substances that exert an injurious effect on the blood-forming organs, the liver or the kidneys

Exposure of the population at large to hazardous environmental pollutants is clearly the most significant danger in terms of numbers that may be affected. However, there is now increasing concern about direct exposure to chemicals in the workplace, where levels may reach much higher concentrations than when widely dispersed outside. As early as 1942, the U.S. Department of Labor suggested that pregnant women should avoid occupational exposure to certain known toxic substances that might be extra hazardous to the foetus (table 2). Discussing this list in her book *Women in industry*, Baetjer wrote in 1946: 'Because these substances may affect adversely the pregnant woman or fetus, the concentrations usually accepted as allowable should not be considered safe for pregnant women.' However, since that time, threshold limit values (t.l.vs) for occupational exposure, recommended by the American Conference of Governmental Industrial Hygienists and largely adopted by U.S. and U.K. regulatory authorities, have been set for less than 500 of the estimated 100000 chemicals in industrial use today (Stellman & Daum 1973). Furthermore, with the possible exception of ionizing radiation, there have been no systematic studies of the relative safety of the accepted t.l.vs with respect to reproductive hazards. Where evidence of adverse effects on male and female sexual function or reproductive capacity does exist, it has been largely ignored in the setting of t.l.vs (Hricko & Marret 1975; Hunt 1977).

It is interesting that t.l.vs in the U.S.S.R. are generally set at much lower levels than in the West since they are based on minimal behavioural effects detectable in exposed animals rather than on more conventional measures of acute toxic effects. The concern for possible reproductive hazards of the workplace in the U.S.S.R. is reflected in a search of the relevant literature with the use of *Index*

Medicus from 1960–76. This revealed 108 relevant papers from Eastern Europe compared with 68 from the rest of the world. It was a Russian survey (Vaisman 1967) which first drew attention to the poor reproductive record of female anaesthetists. Since 1971, the Soviet public health and labour laws have required that special investigations be carried out to determine the effect of industrial and occupational factors on gynaecological and obstetrical morbidity and mortality.

Concern for such issues is, however, increasing in the West; the American College of Obstetricians and Gynecologists, for example, have recently issued a document *Guidelines on pregnancy and work* (A.C.O.G. 1977), which includes discussion of chemical hazards to the pregnant woman from the workplace or the environment. Undoubtedly, interest in this area will be further stimulated by recent legislation. In the U.K., the Employment Protection Act 1975 safeguards the right of the pregnant worker to continued employment during pregnancy. However, it is not difficult to see how such legislation may create considerable problems for an employer who, under the U.K. Congenital Disabilities (Civil Liabilities) Act 1976, may be sued by a disabled child where the disablement is believed to be due to exposure of either of its parents to a toxic chemical in the workplace.

Others, too, have not been slow to point out that equality for women, and health and safety legislation, are powerful movements on a collision course (Rieke 1973). In such a climate, it is particularly disturbing to note the lack of scientific data on which a more rational appraisal of the possible reproductive hazards of work might be based. Proposed legislation in the U.K. envisages very limited testing of new chemicals produced in amounts exceeding 1 t annually, for carcinogenicity, mutagenicity and embryo-toxicity (Health & Safety Commission 1977). However, chemicals already in use will not be subject to these requirements despite the lack of such evidence in the majority of these.

In the remainder of this paper the range of possible adverse effects on reproduction are discussed and such limited data as there are on reproductive outcome after exposure to chemicals in the workplace or the environment are reviewed.

TYPES OF REPRODUCTIVE HAZARD

Exposure during pregnancy

Probably the most widely known example of reproductive toxicity is that due to thalidomide. This drug, if taken during the early stages of pregnancy, almost always produced severe morphological damage to the foetus resulting in the birth of children with a variety of defects. Severe shortening or absence of the limbs was the most obvious of these. The production of gross defects is, however, only one of several types of toxic effect which may result from chemical exposure during pregnancy.

One way of classifying toxic effects on the foetus is in terms of when the effect produced will be recognized. Five major classes of effect can be defined:

1. Immediate effects, such as foetal death and abortion which may occur within a few hours of exposure to a toxic chemical.

2. Effects recognizable at birth: these may range from severe effects such as anencephaly, which are incompatible with extrauterine life, through to the severe but not life-threatening handicaps such as those produced by thalidomide. The milder defects such as cleft lip may be surgically treatable so as to cause no long-lasting impairment.

3. Effects not recognized for a few years: these may range from internal structural defects such as congenital heart disorders which are not detected until the child starts to play active games, through to disorders of central nervous system structure or function. These latter effects may result in behavioural problems or slight reduction in intellectual capacity, or may be severe enough to cause cerebral palsy, mental retardation or sensory defects like blindness.

4. Effects recognized only after many years, such as the transplacental induction of cancer in the daughters of women treated with stilboestrol during pregnancy. These girls appeared quite normal until 15–20 years of age when a small proportion of them developed a rare and sometimes fatal cancer of the vagina. It is not known to what extent other types of cancer may be dependent on prenatal influences, but animal experiments have suggested that the foetus may be especially sensitive to the actions of a wide range of carcinogens.

5. Effects which are not fully recognized until expressed in subsequent generations. It is known from animal experiments that some substances, especially cytotoxic chemicals, may produce a slight or transient reduction in fertility but that the offspring may also have reduced fertility and subsequent generations may be completely infertile (Hemsworth & Jackson 1965).

Exposure of non-pregnant females

A number of industrial chemicals described below have been shown to interfere with the menstrual cycle. This may occur either by an action on the brain to interfere with the release of gonadotrophins or by a direct effect on the ovaries and uterus. In some cases this may result in frequent and excessive bleeding such as that after benzene exposure, which is associated with a poor reproductive history. Menstrual disorders may be an early warning sign of subsequent infertility.

Exposure of males

It is surprising how little interest has been shown in the adverse effects of chemicals on male reproductive capacity. As will be described below, industrial chemicals may affect all aspects of male reproduction: libido; sperm production, morphology and motility; fertility; and mutagenic defects. Since sperm production involves mitotic division of cells throughout the reproductive life of the male, the chance of chemically induced mutations is higher than in the female. Epstein (1972) has reviewed the many diverse effects such as abortion, congenital malformation, genetic disease, decreased life span, infertility, mental retardation, senility and cancer which may follow such mutations.

Specific examples of environmental reproductive toxicity
Effects of chemical exposure in the male

As mentioned above, the possible effects of chemical exposure in the male are wide-ranging, from effects such as diminished libido and potency, which are not only difficult to assess objectively but are so frequently reported in control populations as to be almost impossible to interpret, to problems such as infertility or mutation of the germ cells, which may only be detected by detailed laboratory investigations. With such difficulties surrounding detection, it is not surprising that manufacture of dibromochloropropane (DBCP), a soil pesticide, was only recently discovered to be hazardous by chance when workers in a DBCP plant mentioned to each other that they were having infertility problems. Animal toxicity data, available since 1963, documenting damage to the testes of rats after DBCP exposure and the fact that DBCP is closely related to a well known experimental male antifertility agent, α-chlorohydrin, was apparently ignored until recently.

The effects of lead in the male were first documented as long ago as 1881 by Rennert (see Rom 1976) in pottery glazers, where of 33 children born to normal mothers but where the fathers were exposed to lead, 19 had macrocephaly and 13 suffered from convulsions. In 1905, Rudeaux analysed 442 pregnancies in women married to lead workers, of which 66 ended in abortion and 241 were premature births (Rom 1976). In a study in the 1930s, women whose husbands were exposed to lead had twice as many childless marriages as expected (quoted in Hamilton & Hardy 1974). The data on the possible contribution of chromosome abnormalities in the spermatozoa to adverse pregnancy outcome in wives of lead workers are conflicting. Some investigators have reported an increase in chromosome abnormalities (Schwanitz *et al.* 1970; Forni & Secchi 1972; Lancranjan *et al.* 1975) while others report no differences from controls (O'Riordan & Evans 1974; Bauchinger *et al.* 1972) in cultured blood lymphocyte preparations. What is perhaps more disturbing for pregnancy outcome is the statement from Russian researchers that functional disturbances in children are far more common than mutation or malformations when there is a parental lead burden (Rom 1976).

The list of synthetic and naturally occurring chemical and biological (virus) agents that may be mutagenic (i.e. have demonstrated such potential in mutagenicity testing systems) is enormous. However, detailed consideration has been given to at least one category of workers, those exposed to vinyl chloride monomer (VCM), known to be mutagenic in microbial test systems. Infante *et al.* (1976*a*), summarizing the current literature, cite studies from four different countries showing a significant increase in chromosomal aberrations in lymphocytes of workers occupationally exposed to VCM. If such mutagenic changes can also affect the germ cells, then an increase in intrauterine deaths and congenital malformations among their offspring would be expected. Two studies suggest that this may indeed be the case; Selikoff (1974) in an uncontrolled study reported a stillbirth and miscarriage rate among wives of exposed VCM workers of 7–14 per 100 pregnancies, a rate which Infante (1976b) regards as being high, since it is 2–4 times higher than the rate reported for the State of Georgia for which detailed records exist. This has been confirmed in well controlled studies of workers before and after exposure to VCM (Infante 1976c). Similarly, an increase in congenital malformations, particularly of the central nervous system, has been reported in three Ohio communities with PVC production facilities (Infante 1976b), though its association with direct vinyl chloride exposure of the father is disputed (Edmonds et al. 1975).

There are suggestions of a similar occupational problem with *anaesthetics* which have recently been shown to cause chromosome aberrations in non-mammalian test systems (Grant et al. 1977). Small increases in the risk of spontaneous abortion (Askrog & Harvald 1970) and congenital abnormalities (Cohen et al. 1974; Knill-Jones et al. 1975) have been reported in the wives of male anaesthetists, but more data are needed before the claim can be accepted.

Other miscellaneous chemicals have also been reported to affect the occupationally exposed male. Unsurprisingly, male workers in pharmaceutical plants manufacturing oestrogens have reported impotence. Recent experience in Virginia, America, has documented loss of libido among the numerous symptoms of kepone poisoning developed by over half the work force in a plant manufacturing this pesticide (N.I.O.S.H. 1976). Boric acid is also suspect; in a small study of 28 male workers engaged in boric acid production, questioning revealed loss of libido, and analysis of seminal fluid in six of them disclosed reduced volumes, low sperm counts and elevated fructose contents. These findings correlated with animal experiments where chronic exposure of male rats to boric acid aerosol, in amounts producing insignificant general toxicity, caused sterility within 4 months, suggesting specific gonadal toxicity (Tarasenko et al. 1972). Carbon disulphide has been implicated in the male as well as in the female (see later), with reports of decreased libido, failure of erection and a fivefold increase in sperm abnormalities in men occupationally exposed for 3 years (Lancranjan 1972).

The effects of radiation on fertility have been well documented (see, for example, Kitabatake et al. 1974), yet it is curious that for radiation, lead and carbon disulphide, exposure of the pregnant female is controlled by legislation in many countries including the U.K., while the effects of low-level exposure of the male to many substances are largely ignored. It is important in the development of this area of occupational medicine that concern for direct toxic hazards to the foetus does not override that for the species as a whole, which could result from testicular damage. For example, it is known that in certain situations, the testes are more affected than the relatively protected ovaries and that account should be taken of this in the setting of t.l.vs. This would be particularly important with respect to mutagenic compounds, the effects of which may not be evident for a number of generations.

Effects of chemical exposure in the female

Women currently form a substantial proportion of the workforce and their numbers are still increasing: in the U.K., women are now 36 % of the workforce, in the U.S. 40 % and in Germany 35 %. The extent of employment during pregnancy is also considerable. In 1963, an American survey showed 31 % of 3780 mothers who had legitimate births were employed outside the home at some stage of pregnancy and half of these were still working after the sixth month (Diddle 1970). The proportion of women working during pregnancy is likely to have increased since then, and the range of chemicals and radiation sources to which they may have been exposed is large (see, for example, Kullander *et al.* 1976).

Disorders of menstrual function, pregnancy, the outcome of pregnancy and subsequent development of the child are all pertinent areas for investigation of occupational reproductive hazards to the female. Menstrual disorders, while in themselves having no direct effect on subsequent generations, should be regarded as important indicators of possible hazard, since they may reflect underlying endocrine disturbances of consequence during future pregnancies. The fundamental modification to respiratory and cardiovascular dynamics during pregnancy also raise the possibility of increased susceptibility to occupationally induced disease of the pregnant woman herself, aside from any considerations of the particular vulnerability of the foetus.

Information on spontaneous abortion, stillbirth, prematurity, birth mass, congenital malformation, and perinatal and infant mortality, in relation to maternal and paternal occupation, would provide vital clues as to the extent of reproductive hazards of work. Yet in the U.K. it is only since 1974 that the occupations of both mother and father have been recorded on the Congenital Malformation Notification System and the Stillbirth Register. On the Birth Records themselves, only the father's occupation is recorded. In cases of illegitimate birth, only the mother's occupation is recorded, regardless of outcome. At present very few of the records are coded and analysed, and even if they were, correlating adverse pregnancy outcome with maternal occupation would be very difficult since the total numbers of births to women workers in any specific occupation is not known. In America the situation is even worse since paternal occupation was dropped from Standard Certificates in 1968 because educational attainment was found to be a more reliable indicator of socio-economic status than occupation. Maternal occupation is not recorded. Suggestions for improving current methods of data collection to facilitate appropriate studies in this area have been discussed in detail elsewhere by Hunt (1975).

Metals

Menstrual disorders are said to occur more commonly in women handling lead (Stofen & Waldron 1974), and the ability of lead to cause abortion has long been exploited by women seeking to end an unwanted pregnancy. However, there appear to be no convincing studies in man to show that lead levels below those causing abortion are teratogenic (Clayton 1975), though a correlation has been reported between atmospheric lead concentrations and deaths from congenital malformations (Hickey 1970). The incidental anaemia of lead poisoning (Angle & McIntyre 1964) could also affect the foetus. Lead does cross the placenta and in sheep it has been claimed that prenatal exposure to low lead levels impairs subsequent learning ability in the offspring (E.P.A. 1973).

Lead is also secreted in the breast milk and, in animal experiments, offspring of rats fed lead only during lactation were smaller than normal and had c.n.s. defects including paralysis (N.A.S. 1972). Kostial & Momčilović (1974) have shown that lead transfer in the rat to the offspring is highest during late lactation. If this were so in humans, it could have serious consequences for brain development which proceeds until at least 18 months of age (Dobbing 1974) and possibly considerably longer. In view of this and the known effects of severe lead exposure in young children on subsequent mental development, the present caution in this area would seem well advised.

Occupational exposure to metallic mercury is reportedly associated with a high incidence of abnormal ovarian function (Panova & Ivanova 1976). The use of mercury for the treatment of syphilis in the last century caused many abortions (Afonso & Alvarez 1960), an outcome which may have been best on balance as far as the foetus was concerned. In contrast to organic mercury, which is firmly established as directly embryotoxic in humans (see later), the possible teratogenic effects of inorganic mercury are not known.

Selenium has remained a candidate for suspicion following a single report of a 5 year experience in a laboratory where it was used in culture media. Of eight women exposed, four, possibly five, aborted and one delivered a baby with club foot (Robertson 1970). Cadmium too has received some attention. Tsvetkova (1970) recorded significantly lower birth masses in babies of workers occupationally exposed to cadmium but could find no other obstetric or gynaecological problems. In reviewing the subject, Webb (1975) concluded that the effects of testicular damage, impaired reproduction and teratogenicity were confined to high-level exposure in animals and that it was not a hazard for man. The numerous reports of the embryotoxicity in animals of a large number of metals, summarized elsewhere (Beliles 1975; Wilson 1977), should, however, alert us to the possibility that there may be a number of yet undisclosed hazards associated with occupational exposure to metals.

Beryllium may not be a particular hazard to the foetus but the pregnant woman herself may be particularly vulnerable to beryllium toxicity; in one series an astonishing 63 out of 95 fatal cases of beryllium poisoning in women were in association with pregnancy (Hardy 1965). Beryllium is known to cross the placenta and has been found in the baby's urine (Hall *et al.* 1959), but its prenatal toxicity is unknown.

Gases

Direct exposure of women doctors or nurses to anaesthetic gases has been claimed to result in a higher incidence of infertility, spontaneous abortion, still-birth, low birth mass and congenital malformation, though not all of the findings have been replicated in every study (Vaisman 1967; Askrog & Harvald 1970; Cohen *et al.* 1971; Knill-Jones *et al.* 1972; Corbett *et al.* 1974; Knill-Jones *et al.* 1975; Pharoah *et al.* 1977). Furthermore, the significance and interpretation of some of the findings have been challenged (Rushton 1976) and doubts have been expressed as to how legitimate it is to single out anaesthetic gas exposure as the causal factor when there are clearly other adverse factors at work in operating theatres, such as stress (Smithells 1976). However, after a nationwide survey by the American Society of Anesthesiologists sponsored by N.I.O.S.H. (Cohen *et al.* 1974) covering 50000 operating-room personnel, the findings were regarded as serious enough to warrant a decision to exclude all pregnant doctors and nurses from operating areas for a trial period of a few years. In the U.K., where such decisions are invariably taken at a slower pace, a Department of Health circular has been issued drawing attention to the possible hazard, advocating that staff be fully informed of its nature and that pollution levels in operating theatres be reduced (D.H.S.S. 1976).

The ability of another gas, carbon monoxide, to lower the oxygen-carrying capacity of the blood, has for a long time placed it under suspicion. In the few recorded cases of acute poisoning of the mother, foetal death or severe brain damage have usually been the outcome (Longo 1970), but the effect of low levels of carbon monoxide such as might be encountered in occupational exposure to automobile exhaust is uncertain. The literature on the adverse effects of smoking during pregnancy, including possible late effects on child development (Butler & Goldstein 1973), is now large. Smokers have higher carboxyhaemoglobin levels than non-smokers, but it is not known whether the adverse effects are due to this and/or the other toxic components of tobacco smoke. However, it is worth noting that mean carboxyhaemoglobin levels in mothers who smoke during pregnancy range from 2.0 to 8.3 % (several studies summarized by Longo 1970) while occupational exposure to automobile exhaust has produced levels of up to 10 % (Goldsmith 1970).

Miscellaneous organic chemicals

A number of miscellaneous industrial chemicals are claimed to affect reproductive function adversely, but few of the reports specify the dose at which the effects were found, and in general the published details are inadequate to assess the validity of the claims. An increased incidence of anovulatory cycles with low oestrogen levels in women working with phthalate plasticizers is associated with a poor prognosis for the outcome of pregnancy, principally spontaneous abortion (Aldyreva *et al.* 1975). A high incidence of intrauterine death has also been found in experimental animals exposed to these plasticizers (Aldyreva *et al.* 1975).

Russian studies have also shown that women working with formaldehyde are 2.5 times more prone than controls to develop menstrual disorders (dysmenorrhoea), more likely to develop toxaemia or anaemia during pregnancy, and more likely to have spontaneous abortions. In those carrying to term, the incidence of babies of low birth mass (defined as 2.5–3.0 kg), was 27 % compared with only 11 % in controls (Shumlina 1975).

Women employed in the textile industry may also be at risk. Carbon disulphide, used extensively in the manufacture of viscose rayon, causes menstrual irregularities, principally prolonged or heavy bleeding and possibly amenorrhoea (Wiley *et al.* 1936), and is further alleged to decrease fertility and increase spontaneous abortions (Ehrhardt 1967). Female workers in the chemical and spinning areas for capron silk show disturbances of menstrual and childbearing functions, caprolactam being the chemical under suspicion (Martynova *et al.* 1972).

Occupational exposure to organochlorine pesticides for varying periods of 1–10 years has been studied in detail in 30 women (Blekherman & Ilyina 1973). Instead of showing the normal cyclical pattern of changes, oestrogen levels remained fairly constant during the menstrual cycle, with no distinct ovulatory peaks of oestrogen or of luteinizing hormone, and a shortened luteal phase. The authors concluded from this study, in which a number of both adrenal and ovarian parameters were followed, that oestrogen levels were the most sensitive indicator of ovarian malfunction. Veis (1970) has shown an increased incidence of toxaemia among organochlorine pesticide production workers during pregnancy.

Women involved in polystyrene processing have been found to be more prone to irregular menstrual cycles and toxaemia of pregnancy, while experiments on animals have confirmed that there is suppression of pituitary–gonadal function after exposure to low levels of styrene (Zlobina *et al.* 1975).

Contact with the organic solvents benzene, toluene and xylene is reported to cause prolonged and heavy menstrual bleeding (*N.I.O.S.H. criteria document of toluene* 1973; Syrovadko 1973), which may contribute to the anaemia caused directly by benzene. Some reports suggest that pregnancy during or after benzene exposure may precipitate aplastic anaemia (Browning 1965).

Industrial chemicals can also be transmitted in the breast milk and may change its quantity and composition. Occupational exposure to the combination of gasoline and chlorinated hydrocarbons, the latter being found in the milk, also reduced lactation (Mukhametova & Vozovaya 1972). Chloroprene resin workers have been found to have altered amino acid content in their breast milk (Vanuni 1974).

With the slowly increasing return to popularity of breast feeding, particularly among the poorer industrial communities (Lillington 1975; Jepson *et al.* 1976) and the increasing proportion of mothers returning to work (Diddle 1970), this area clearly requires more research.

Environmental contaminants

Methyl mercury is the one metallic compound firmly established as directly embryotoxic in humans. Industrial pollution of the water in Minamata Bay in the 1950s and later at Niigata in Japan resulted in cerebral palsy and mental retardation in the offspring of pregnant women who consumed fish contaminated with methyl mercury, 6 % of all births being affected in the Minamata disaster. Subsequent isolated cases of mercury ingestion by pregnant women and a further epidemic of poisoning in Iraq in 1972, where grain treated with methyl mercury was used to make bread, have confirmed its toxicity to the central nervous system, affected infants being born to apparently symptomless mothers (reviewed by Koos & Longo 1976). The ability of methyl mercury to concentrate in the foetal brain, some would claim in higher quantities than that found in the maternal brain (Clarkson 1972), may explain its potency as a c.n.s. teratogen.

There is little doubt that methyl mercury can also be transmitted in the breast milk. Blood mercury levels of infants born before the epidemic of poisoning in Iraq (i.e. not exposed *in utero*) were as high as maternal values in some cases, and breast milk was shown to contain 3–6 % of maternal blood concentrations (Bakir *et al.* 1973). It seems likely that in some cases the neurological deficits induced by prenatal exposure may have been exacerbated by continued postnatal exposure through suckling.

The recent explosion at the chemical factory in Seveso, Italy, where the highly poisonous substance dioxin was dispersed over a wide area of the surrounding villages, again focused attention on environmental hazards of industry. Reports of an increased incidence of abortion, stillbirth and malformations from Vietnam (N.R.C. 1974), where there was widespread spraying of the countryside with 2,4,5-T, a herbicide containing dioxin as a contaminant, have been received with some scepticism in the West. Nevertheless, the potency of dioxin as a teratogen in animals (Neubert & Dillmann 1972; Courtney & Moore 1971) has caused widespread concern for the outcome of pregnancy in women living in the Seveso area. We do not yet know whether this concern will turn out to be justified.

In Japan, maternal ingestion of polychlorinated biphenyls (PCBs) from contaminated rice oil in 1968 has given rise to ten cases of small-for-dates babies, with abnormal cola-coloured skin and ocular discharge, some affected babies being born to symptomless mothers (Kuratsune *et al.* 1972). The colour has faded during infancy but later effects, including early tooth eruption and gingival hyperplasia in babies exposed *in utero* have been found (Fraumeni 1974), while exposure through breast milk may also have long-term effects. In Japan, follow-up of children who received PCBs through the breast milk only, showed hypotonia and apathy and the children seemed 'sullen and expressionless' (Miller 1977), the abnormalities persisting up to 6 years of age in some cases. This has raised suspicions about occupational exposure to low levels of PCBs but we have been unable to find any reports of research done in this area. The possible effects of polybrominated biphenyls (PBBs) have been the subject of recent speculation (Finberg 1977; Anon. 1977) after the large-scale accidental contamination of meat and dairy products in Michigan, U.S.A. Some reports record disturbances in the oestrous cycles and an increased incidence of stillbirths in cattle consuming contaminated feeds (Mercer *et al.* 1976; Jackson & Halbert 1974) but adverse effects on pregnancy in humans have not been reported.

Once into the food chain, pesticides such as DDT are almost impossible to eliminate. The effect of constant background exposure to low levels of such substances are unknown, and are hard to investigate in view of the difficulty of finding pesticide-free control populations. Transmission of chlorinated hydrocarbons including pesticides in the breast milk (Yeh *et al.* 1976; Bakken & Seip 1976) should perhaps come under particularly close scrutiny, since fat in the breast milk is one of the principal routes by which persistent chemicals like DDT, BHC, PCBs and PBBs may be cleared from the body (Finberg 1977; Miller 1977). Although residual levels of DDT may be falling in countries such as America, Japan and England after banning of widespread agricultural use, other pesticides such as BHC are now being found in higher quantities than DDT in the breast milk (Yeh *et al.* 1976).

Transplacental carcinogenesis

An increased incidence of leukaemia and other cancers found by MacMahon (1962) and Stewart & Kneale (1970a, b) in children irradiated *in utero* has not always been confirmed elsewhere (B.E.I.R. Report 1972). However, the existence of such late effects on the offspring is now fairly well established with respect to prenatal exposure to the drug diethylstilboestrol (DES) (Herbst *et al.* 1974) which causes a rare form of vaginal cancer, only emerging around puberty in daughters of mothers given DES to prevent miscarriage. Although DES has not been shown to cause cancer in male offspring of exposed mothers, other disturbing late effects have now emerged: in a study of 163 male offspring, 25% had epididymal cysts, hypotrophic testes, and low semen volumes with 'severely pathologic' sperm quality scores, while two were azoospermic (Bibbo *et al.* 1977).

The possibility of transplacental carcinogenesis has also been raised with anaesthetics; Corbett *et al.* (1974) found three neoplasms in 2 children of 434 born to mothers who were practising anaesthetists during pregnancy. One case of leukaemia was reported from the 261 births to anaesthetists not practising during pregnancy. As one of the authors points out, these figures are not statistically significant. However (Corbett 1976), in view of the reports of increased incidence of cancer in adult anaesthetists (Bruce *et al.* 1968; Corbett *et al.* 1974; Cohen *et al.* 1974) and the likelihood that some substances that are carcinogens in adults will also be carcinogenic in the foetus, these studies seem worth following up. Another disquieting finding relates to vinyl chloride, a probable carcinogen in adults (Infante 1976*b*); Maltoni & Lefemine (1974) have reported angiosarcomas of the subcutaneous tissues in offspring of pregnant rats exposed to VCM; the neoplasm was similar to that found in adults (Monson *et al.* 1974).

Paternal exposure to hydrocarbons has also been suggested as a cause of cancer in the offspring. A Quebec study of 386 children dying from malignant disease before 5 years of age showed a significant excess of fathers in hydrocarbon-related occupations with relative odds of 2:1 (Fabia & Thuy 1974). However, a recent search of the Finnish Cancer Registry failed to confirm any such association (Hakulinen *et al.* 1976).

In view of the fact that cancer, much of it congenitally acquired, accounts for a significant proportion of childhood deaths, the possible influence of maternal and paternal occupation warrants further study, though again the contribution of any one industrial factor is likely to be small and therefore very difficult to detect.

Conclusions

In conclusion it would seem that in the male, environmental chemicals can affect reproductive capacity, and may be associated with an increase in abortions, congenital malformations and childhood cancer in their offspring. In females, environmental chemicals may be more toxic during pregnancy to the mother as well as to the foetus. They may affect fertility and pregnancy, and may be associated with congenital malformations, stillbirths, and abnormal postnatal development. Children may be exposed to environmental chemicals directly by pollution, or indirectly via the parents, which may result in death, neurotoxicity or cancer.

However, from the survey of the literature we have undertaken, one thing emerges most clearly, that there is remarkably little information on the reproductive hazards of the working environment. In this respect it is hoped that recently introduced methods of data collection in the U.K. will provide more information on the relation of maternal or paternal occupation to pregnancy outcome.

References (Sullivan & Barlow)

A.C.O.G. 1977 *Guidelines on pregnancy and work.* American College of Obstetricians & Gynecologists, Chicago, Illinois.

Afonso, J. & De Alvarez, R. 1960 Effects of mercury on human gestation. *Am. J. Obstet. Gynec.* **80**, 145–154.

Aldyreva, M. V., Klimona, T. S., Izyumova, A. S. & Timofievskaya, L. A. 1975 The influence of phthalate plasticisers on the generative function. *Gig. Truda prof. Zabol.* **19**, 25–29.

Angle, C. R. & McIntyre, M. S. 1964 Lead poisoning during pregnancy. *Am. J. Dis. Child.* **108**, 436–439.

Anon. 1977 Editorial: polybrominated biphenyls, polychlorinated biphenyls, pentachlorophenyl – and all that. *Lancet* ii, 19–21.

Askrog, V. F. & Harvald, B. 1970 Teratogenic effect of inhalation anaesthetics. *Nord. Med.* **83**, 498–500.

Bakir, F., Damluji, S., Amin-Zaki, L., Murtadha, M., Khalidi, A., Al-Rawi, N., Tikriti, S., Dhahir, H., Clarkson, T., Smith, J. & Doherty, R. 1973 Methylmercury poisoning in Iraq. An inter-university report. *Science, N.Y.* **181**, 230–241.

Bakken, A. F. & Seip, M. 1976 Insecticides in human breast milk. *Acta paediat. scand.* **65**, 535–539.

Baetjer, A. M. 1946 *Women in industry*. Philadelphia and London: W. B. Saunders.

Bauchinger, M., Schmid, E. & Schmidt, D. 1972 Chromosomen-analyse bei Verkehrspolizisten mit Erhöhter Bleilast. *Mutat. Res.* **16**, 407–412.

B.E.I.R. Report 1972 *The effects on populations of exposure to low levels of ionising radiation*. Washington, D.C.: Biological Effects of Ionising Radiation Advisory Committee, National Academy of Sciences, National Research Council.

Beliles, R. P. 1975 Metals. In *Toxicology, the basic science of poisons* (ed. L. J. Casarett & J. Doull), pp. 461–463. New York: Macmillan.

Bibbo, M., Gill, W. B., Azizi, F., Blough, R., Fang, V. S., Rosenfield, R. L., Schumacher, G. F. B., Sleeper, K., Sonek, M. G. & Wied, G. L. 1977 Follow-up study of male and female offspring of DES-exposed mothers. *Obstet. Gynec., N.Y.* **49**, 1–8.

Blekherman, N. A. & Ilyina, V. I. 1973 Changes of ovary function in women in contact with organochlorine compounds. *Pediatriya* **52**, 57–59.

Browning, E. 1965 *Toxicology and metabolism of industrial solvents*, pp. 27–28. Amsterdam: Elsevier.

Bruce, D. L., Eide, K. A., Linde, H. W. & Eckenhoff, J. E. 1968 Causes of death among anesthesiologists: a 20-year survey. *Anesthesiology* **29**, 565–569.

Butler, N. R. & Goldstein, H. 1973 Smoking in pregnancy and subsequent child development. *Br. med. J.* iv, 573–575.

Clarkson, T. W. 1972 Recent advances in the toxicology of mercury with emphasis on the alkylmercurials. *Crit. Rev. Toxicol.* **1**, 203–234.

Clayton, B. E. 1975 Lead: the relation of environment and experimental work. *Br. med. Bull.* **31**, 237–240.

Cohen, E. N., Belville, J. W. & Brown, B. W. 1971 Anesthesia, pregnancy and miscarriage: a study of operating room nurses and anesthetists. *Anesthesiology* **35**, 343–347.

Cohen, E. N., Brown, B. W., Bruce, D. L., Cascorbi, H. F., Corbett, T. G., Jones, T. W. & Whitcher, C. E. 1974 Occupational disease among operating room personnel: a national study. *Anesthesiology* **41**, 317–340.

Corbett, T. H., Cornell, R. G., Lieding, K. & Endres, J. L. 1974 Birth defects of children among Michigan nurse-anesthetists. *Anesthesiology* **41**, 341–344.

Corbett, T. H. 1976 O.R. exposure – cancer, miscarriages and birth defects. *Women & Hlth* **1** (5), 11–15.

Courtney, K. D. & Moore, J. A. 1971 Teratology studies with 2,4,5-trichlorophenoxyacetic acid and 2,3,7,8-tetrachlorodibenzo-*p*-dioxin. *Toxic. appl. Pharmac.* **20**, 396–403.

D.H.S.S. 1976 *Department of Health and Social Security Health Circular* HC (76) 38, 1976.

Diddle, A. W. 1970 Fetal and maternal morbidity, employment policy and medicolegal aspects. *J. occup. Med.* **12**, 10–15.

Dobbing, J. 1974 The later development of the brain and its vulnerability. In *Scientific foundations of paediatrics* (ed. J. A. Davis & J. Dobbing), pp. 565–577. London: William Heinemann Medical Books.

Edmonds, L. D., Falk, H. & Nissim, J. E. 1975 Congenital malformations and vinyl chloride. *Lancet* ii, 1098.

Ehrhardt, W. 1967 Experience with the employment of women exposed to carbon disulphide. In *International Symposium on Toxicology of Carbon Disulphide*, Prague, 1966. Amsterdam: Excerpta Medica Foundation.

E.P.A. 1973 *Health implications of airborne lead*. U.S. Environmental Protection Agency, November 28, 1973.

Epstein, S. S. 1972 Environmental pathology: a review. *Am. J. Path.* **66**, 352–373.

Fabia, J. & Thuy, T. D. 1974 Occupation of father at time of birth of children dying of malignant diseases. *Br. J. prev. soc. Med.* **28**, 98–100.

Finberg, L. 1977 PBBs: the ladies' milk is not for burning. *J. Pediat.* **90**, 511–512.

Forni, A. & Secchi, G. C. 1972 Chromosome changes in preclinical and clinical lead poisoning and correlation with biochemical findings. In *Proc. of E.P.A. Int. Symp. on Environmental Health Aspects of Lead*, 2–6 October 1972, Amsterdam, pp. 473–485.

Fraumeni, J. F. 1974 Chemicals in human teratogenesis and transplacental carcinogenesis. *Pediatrics* (suppl.) **53**, 807–812.

Goldsmith, J. R. 1970 Contribution of motor vehicle exhaust, industry and cigarette smoking to community carbon monoxide exposures. *Ann. N.Y. Acad. Sci.* **174**, 122–134.

Grant, C. J., Powell, J. N. & Radford, S. G. 1977 The induction of chromosomal abnormalities by inhalational anaesthetics. *Mutat. Res.* **46**, 177–184.

Hakulinen, T., Salonen, T. & Teppo, L. 1976 Cancer in the offspring of fathers in hydrocarbon related occupations. *Br. J. prev. soc. Med.* **30**, 138–140.

Hall, J. C., Wood, C. H., Stoeckle, J. D. & Tepper, L. P. 1959 Case data from the beryllium registry. *A.M.A. Arch. ind. Hyg.* **19**, 100–103.

Hamilton, A. & Hardy, H. L. 1974 *Industrial toxicology*, 3rd rev. edn. New York: Publishing Science.

Hardy, H. L. 1965 Beryllium poisoning. Lessons in control of man-made disease. *New Engl. J. Med.* **273**, 1188–1199.

Health & Safety Commission 1977 *Proposed scheme for the notification of the toxic properties of substances: discussion document.* London: H.M.S.O.

Hemsworth, B. N. & Jackson, H. 1965 Embryopathies induced by cytotoxic substances. In *Embryopathic activity of drugs* (ed. J. M. Robson, F. M. Sullivan & R. L. Smith), pp. 116–137. London: Churchill.

Herbst, A. L., Robboy, S. J., Scully, R. E. & Poskanzer, D. C. 1974 Clear-cell adenocarcinoma of the vagina and cervix in girls: analysis of 170 registry cases. *Am. J. Obstet. Gynec.* **119**, 713–724.

Hickey, R. J. 1970 Ecological statistical studies concerning environmental pollution and chronic disease. In *Digest of technical papers, 2nd Internat. Geo. Science Electronics Symposium*, Washington, D.C., 14–17 April 1970, p. 13.

Hricko, A. M. & Marrett, C. B. 1975 *Women's occupational health: the rise and fall of a research issue.* New York: American Association for the Advancement of Science.

Hunt, V. R. 1975 *Occupational health problems of pregnant women: a report and recommendations for the Office of the Secretary, Department of Health, Education and Welfare.*

Hunt, V. R. 1979 Protection of workers' health. In *Proceedings of a Conference on Protective legislation and women's jobs: reevaluating the past and planning for the future*, Smith College, Mass., Nov. 1977. (In the press.)

Infante, P. F., McMichael, A. J., Wagoner, J. K., Waxweiler, R. J. & Falk, H. 1976*a* Genetic risks of vinyl chloride. *Lancet* i, 734–735.

Infante, P. F. 1976*b* Oncogenic and mutagenic risks in communities with polyvinyl chloride production facilities. *Ann. N.Y. Acad. Sci.* **271**, 49–57.

Infante, P. F. 1976*c* Carcinogenic, mutagenic and teratogenic risks associated with vinyl chloride. *Mutat. Res.* **41**, 131–141.

Jackson, T. F. & Halbert, F. L. 1974 A toxic syndrome associated with the feeding of polybrominated biphenyl-contaminated protein concentrate to dairy cattle. *J. Am. vet. med. Ass.* **165**, 437–439.

Jepson, M. E., Smith, B. A. M., Pursall, E. W. & Emery, J. L. 1976 Breast-feeding in Sheffield. *Lancet* ii, 425–426.

Kitabatake, T., Watanabe, T. & Sato, T. 1974 Sterility in Japanese radiological technicians. *Tohoku J. exp. Med.* **112**, 209–212.

Knill-Jones, R. P., Moir, D. D., Rodrigues, L. V. & Spence, A. A. 1972 Anaesthetic practice and pregnancy: controlled survey of women anaesthetists in the United Kingdom. *Lancet* i, 1326–1328.

Knill-Jones, R. P., Newman, B. J. & Spence, A. A. 1975 Anaesthetic practice and pregnancy: controlled survey of male anaesthetists in the United Kingdom. *Lancet* ii, 807–809.

Koos, B. J. & Longo, L. D. 1976 Mercury toxicity in the pregnant woman, fetus and newborn infant. *Am. J. Obstet. Gynec.* **126**, 390–409.

Kostial, K. & Momčilović, B. 1974 Transport of lead 203 and calcium 47 from mother to offspring. *Archs envir. Hlth* **29**, 28–30.

Kullander, S., Källen, B. & Sandahl, B. 1976 Exposure to drugs and other possibly harmful factors during the first trimester of pregnancy: comparison of two prospective studies performed in Sweden 10 years apart. *Acta obstet. gynec. scand., Stoch.* **55**, 395–405.

Kuratsune, M., Yoshimura, T., Matsuzaka, J. & Yamaguchi, A. 1972 Epidemiologic study on Yusho, a poisoning caused by ingestion of rice-oil contaminated with commercial brand of polychlorinated biphenyls. *Environ. Hlth Perspect.* **1**, 119.

Lancranjan, I. 1972 Alterations of spermatic liquid in patients chronically poisoned by carbon disulphide. *Medna Lav.* **63**, 29–33.

Lancranjan, I., Posecu, H., Galvenescu, O. *et al.* 1975 Reproductive ability of workmen occupationally exposed to lead. *Archs envir. Hlth* **30**, 396–401.

Lillington, A. W. 1975 Inflation hits the bottle. *Lancet* ii, 512–513.

Longo, L. D. 1970 Carbon monoxide in the pregnant mother and fetus and its exchange across the placenta. *Ann. N.Y. Acad. Sci.* **174**, 313–341.

MacMahon, B. 1962 Prenatal X-ray exposure and childhood cancer. *J. natn. Cancer Inst.* **28**, 1173–1191.

Maltoni, C. & Lefemine, G. 1974 Carcinogenicity bioassays of vinyl chloride. 1. Research plan and early results. *Environ. Res.* **7**, 387–405.

Martynova, A. P., Lotis, V. M., Khadzieva, E. D. & Gaidova, E. S. 1972 Occupational hygiene of women engaged in the production of capron (6-handecanone) fiber. *Gig. Truda prof. Zabol.* **16** (11), 9–13.

Mercer, H. D., Teske, R. H., Condon, R. J., Furr, A., Meerdink, G., Buck, W. & Fries, G. 1976 Herd health status of animals exposed to polybrominated biphenyls (PBB). *J. Toxic envir. Hlth* **2**, 335–349.

Miller, R. W. 1977 Pollutants in breast milk. *J. Pediat.* **90**, 510–511.

Monson, R. R., Peters, J. M. & Johnson, M. N. 1974 Proportional mortality among vinyl-chloride workers. *Lancet* ii, 397–398.

Mukhametova, G. M. & Vozovoya, M. A. 1972 Reproductive power and incidence of gynecological affections among female workers exposed to a combined effect of gasoline and chlorinated hydrocarbons. *Gig. Truda prof. Zabol.* **16** (11), 6–9.

N.A.S. 1972 *Lead: airborne lead in perspective*, p. 162. National Academy of Sciences, Washington, D.C.: Committee on Biological Effects of Atmospheric Pollutants.

Neubert, D. & Dillman, L. 1972 Embryotoxic effects in mice treated with 2,4,5-trichloro-phenoxyacetic acid and 2,3,7,8-tetrachlorodibenzo-*p*-dioxin. *Naunyn-Schmeidebergs Arch. exp. Path. Pharmak.* **272**, 243–264.

N.I.O.S.H. 1973 *Criteria Document of Toluene*, p. 19. National Institute for Occupational Safety and Health.

N.I.O.S.H. 1976 Memorandum from the Director of N.I.O.S.H. to Assistant Secretary for Health on Recommended Kepone Standard, 27 January 1976.

N.R.C. 1974 *The effects of herbicides in South Vietnam. Part A. Summary and conclusions prepared for Department of Defense*. Washington, D.C.: National Research Council.

O'Riordan, M. L. & Evans, H. J. 1974 Absence of significant chromosome damage in males occupationally exposed to lead. *Nature, Lond.* **247**, 50–52.

Panova, S. & Ivanova, S. 1976 Changes in the ovarian function and some functional indices of liver of women in professional contact with metallic mercury. *Akush. Ginekol.* **15** (2), 136–137.

Pharoah, P. O. D., Alberman, E., Doyle, P. & Chamberlain, G. 1977 Outcome of pregnancy among women in anaesthetic practice. *Lancet* i, 34–36.

Rieke, F. E. 1973 Thirty-two million women at work – how different are they? *J. occup. Med.* **15**, 729–732.

Robertson, D. S. E. 1970 Selenium – a possible teratogen. *Lancet* i, 518–519.

Rom, W. N. 1976 Effects of lead on the female and reproduction: a review. *Mt Sinai J. Med.* **43**, 542–552.

Rushton, D. I. 1976 Anaesthetics and abortions. *Lancet* ii, 147.

Schwanitz, G., Lehnert, G. & Gebhart, E. 1970 Chromosome damage after occupational exposure to lead. *Dt. med. Wschr.* **95**, 1636–1641.

Selikoff, I. J. 1974 *N.I.E.H.S. Conference on Public Health Implications of Plastics Manufacture*, Pinehurst, North Carolina, July 1974.

Shumlina, A. V. 1975 Menstrual and child-bearing functions of female workers occupationally exposed to the effects of formaldehyde. *Gig. Truda prof. Zabol.* **19** (12), 18–21.

Smithells, R. W. 1976 Environmental teratogens of man. *Br. med. Bull.* **32**, 27–33.

Stellman, J. M. & Daum, S. M. 1973 *Work is dangerous to your health.* New York: Vintage Books.

Stewart, A. & Kneale, G. W. 1970a Radiation dose effects in relation to obstetric X-rays and childhood cancers. *Lancet* i, 1185–1188.

Stewart, A. & Kneale, G. W. 1970b Age-distribution of cancers caused by obstetric X-rays and their relevance to cancer latent periods. *Lancet* ii, 4–8.

Stofen, D. & Waldron, H. A. 1974 *Sub-clinical lead poisoning.* New York: Academic Press.

Syrovadko, O. N. 1973 Effect of working conditions on the health and some specific functions in female workers exposed to white spirit. *Gig. Truda prof. Zabol.* **17** (6), 5–8.

Tarasenko, N. Y., Kasparov, A. A. & Strongina, O. M. 1972 The effect of boric acid on the generative function in males. *Gig. Truda prof. Zabol.* **16** (11), 13–16.

Tsvetkova, R. P. 1970 Influence of cadmium compounds on the generative function. *Gig. Truda prof. Zabol.* **14** (3), 31–33.

U.S. Department of Labor 1942 Standards for maternity care and employment of mothers in industry. *J. Am. med. Ass.* **120**, 55–56.

Vaisman, A. I. 1967 Working conditions in surgery and their effect on the health of anaesthesiologists. *Éksp. Khir. Anesteziol.* **3**, 44–49.

Vanuni, S. O. 1974 Comparative characteristics of individual and total amino acids in breast milk of parturients: workers and women living in the villages located at different distances from the synthetic chloroprene resin plant. *Zh. éksp. klin. Med.* **14**, 96–101.

Veis, V. P. 1970 Some data on the status of the sexual sphere in women who have been in contact with organochlorine compounds. *Pediat. Akush. Ginek.* **32** (1), 48–49.

Webb, M. 1975 Cadmium. *Br. med. Bull.* **31**, 246–250.

Wiley, F. H., Hueper, W. C. & Von Oettingen, W. F. 1936 On toxic effects of low concentrations of carbon disulfide. *J. ind. Hyg. Toxicol.* **18**, 733–740.

Wilson, J. G. 1977 Environmental chemicals. In *Handbook of teratology*, vol. 1 (Principles and etiology) (ed. J. G. Wilson & F. C. Fraser), pp. 357–368. New York and London: Plenum Press.

Yeh, C. Y., Kuo, P. H., Tsai, S. T., Wang, G. Y. & Wang, Y. T. 1976 A study of pesticide residues in umbilical cord blood and maternal milk. *J. Formosan med. Ass.* **75**, 463–470.

Zlobina, N. S., Izyumova, A. S. & Ragulie, N. Y. 1975 The effect of low styrene concentrations on the specific functions of the female organism. *Gig. Truda prof. Zabol.* **19** (12), 21–25.

Discussion

D. G. WIBBERLEY (*Department of Pharmacy, University of Aston, Gosta Green, Birmingham B4 7ET, U.K.*). My colleagues and I in Birmingham have been using analytical chemistry as a tool for investigating malformations and abortions. What we have done is to examine placental samples collected by Professor J. Edwards from the Birmingham Maternity Hospital over the last 7 years for a range of environmental substances that could be teratogenic. We have determined levels of diethylhexylphthalate, selenium, cadmium and lead. With lead, placental levels showed a wide variation even in normal births (0.3–3.0 parts/10^6) but average levels were about 0.8 parts/10^6, whereas placental lead levels where stillbirth or neonatal death occurred were roughly twice normal levels (1.4–1.7 parts/10^6). The fact that levels tended to be high for most such births, even those with chromosomal abnormalities, leads us to suggest that lead tends to accumulate in general conditions of foetal distress. Further work on stillbirth tissues might tell us whether, in addition, high exposure to lead is having a direct deleterious effect.

SIR RICHARD DOLL (13 *Norham Gardens, Oxford OX2 6PS, U.K.*). The occupations of the fathers of all children who die under 15 years of age are recorded on British death certificates and G. J. Draper & B. M. Sanders (personal communication) have used this information to calculate the proportional mortality attributable to cancer for children of fathers in the main occupational groups recorded by the Registrar-General. No exact comparison with the Canadian data on chemical workers is possible, but the nearest similar groupings gave proportional mortalities attributable to cancer of less than 100 % round both the 1961 and 1971 censuses.

K. P. DUNCAN (*Health and Safety Executive, Baynards House*, 1 *Chepstow Place, Westbourne Grove, London W2 4TF, U.K.*). There are several quantitative biological points that I should like to take up, but for brevity I confine myself to three semi-administrative ones.

(1) No one in the health and safety business in the U.K. could accept Dr Sullivan's view of 'threshold limit' (not limiting) values as if it were a dividing line with his 'satisfactory below and bad above' concept.

(2) We have not ignored the special risks to the foetus and consequently the need to review, for example, lead exposures of women. Various steps have been agreed by a joint H.S.E./M.R.C. Working Group and will shortly be incorporated in new exposure standards for lead workers.

(3) It should not be forgotten that many so-called protective measures for women workers have no biological basis. Jointly with the Equal Opportunities Commission we are reassessing this whole field to set it on a more scientific and less traditional male chauvinist basis.

G. KAZANTZIS (*Department of Community Medicine, The Middlesex Hospital, London W1, U.K.*). Dr Sullivan referred to lead, mercury, cadmium, selenium and beryllium as metals suspected of being potential human teratogenic agents, using this term in its widest sense to include not only malformation but also effects on the subsequent development and behaviour of offspring. However, it is only with certain alkyl mercury compounds, in particular with methyl mercury, that we have any direct evidence of such an effect in man. Environmental contamination of Minamata Bay in Japan with methyl mercury was followed by the birth of a number of babies with a neurological disorder resembling cerebral palsy, with severe physical and mental impairment. This observation was confirmed after the large-scale eating of methyl mercury contaminated bread in Iraq in 1972; in some cases the defect was not apparent until some time after birth (Skerfvius, S. B. & Copplestone, J. F. (1976) *Bull. W.H.O.* **54**, 101–112). Raised levels of mercury were found in the mothers, some of whom did not show clinical evidence of methyl mercury poisoning, and even higher levels were found in their affected offspring. It is known that methyl mercury readily traverses the placental barrier and has a higher affinity for foetal than for adult haemoglobin. The clinical phenomena can therefore be explained by the known behaviour of the metal in the organism. We do not have any data that are at all comparable for the other metals.

ANNE MCLAREN (*M.R.C. Mammalian Development Unit, Wolfson House, 4 Stephenson Way, London NW1 2HE, U.K.*). The problem of testing chemicals for teratogenicity is one that exercises scientists, industry, and governmental agencies. A good correlation has been established between mutagenicity and carcinogenicity, so that if a new chemical substance scores as mutagenic on, for example, the Ames test, the chances are it will turn out to be carcinogenic too. But neither mutagenicity nor carcinogenicity is at all well correlated with teratogenicity. This means that tests for teratogenicity at present have to be carried out *in vivo*, on pregnant mammals. Even then there are many problems, because the relation of mother to foetus is different in every mammalian species, and no good animal model for human pregnancy has yet been found. Bearing in mind that it is important not only to minimize possible risks to the foetus, but also to avoid delaying too long the release of potentially useful chemicals and drugs, does Dr Sullivan think that the guidelines for teratogenicity testing in this country strike a good balance?

F. M. SULLIVAN. I think that the guidelines for reproduction tests for potential new drugs are reasonably satisfactory at present, and in some respects are in advance of those in other countries. For example, Britain was the first country to introduce a requirement in the tests for evidence of effects on special senses and behaviour. New guidelines are currently being drawn up for reproduction tests for pesticides and similar agricultural chemicals. Personally, I feel that in Britain a reasonably good balance is maintained which recognizes the limitations of animal models. Quite often, one is able to analyse the mode of action of teratogenic effects observed in animals and this helps in the extrapolation from animal experiments to man.

For other chemicals covered by the Health and Safety at Work Act, the original proposals in their discussion document for an 'embryolethality' test as a substitute for a teratogenicity test was clearly not satisfactory. I am not sure how the problem of devising a cheap, rapid teratogenicity screen can be solved.

Closing remarks by the chairman of the session on mutagenesis and teratogenesis

The hazards of environmental chemicals, not just to ourselves, to our own generation, but to our children, to posterity, is a topic of vital concern to scientists and non-scientists alike. We as a community might settle for a certain level of risk to ourselves, whether for personal gratification (smoking) or for social gain (use of seed dressings to increase food supply); but to what extent are we thereby putting future generations at risk?

Mutagenesis, the induction of heritable change, has been intensively studied by geneticists for several decades. We know now that some agents (radiation, certain types of chemical) are likely to affect the genetic material of any organism, whether mammal, insect, bacterium or plant. On this basis it has been possible to devise screening tests for potentially mutagenic substances, using *Drosophila* or even bacteria, with results that have some predictive power for our own species. The value of such tests is of course restricted by species specificity for such factors as, for example, metabolic pathways of activation and detoxification. How much genetic disease we may be bequeathing to our descendants, as a result of induced mutations, has been discussed by Professor Vogel.

The topic of teratogenesis, reviewed by Dr Sullivan, concerns abnormalities caused by environmental agents acting before birth. The damage is not genetic: only one generation is involved, but it is an important one. In the past few years, some teratogenic agents have been identified (thalidomide, rubella) because they caused large effects, localized in time and space. Such disasters of course catch the public eye. But when we remember that less than 5 % even of congenital malformations, let alone more intangible deleterious effects, can be attributed to any known teratogenic agent, we can hardly doubt that there are other agents in the environment, as yet unidentified, perhaps with effects still more drastic than thalidomide or rubella virus. Teratogenesis is a very emotive area, and understandably so; it is also one where solid data are particularly hard to obtain, because of the very complex relation between the maternal organism and the unborn young.

I should like to conclude the discussion by stressing the crucial importance of basic research on human pregnancy and the developmental mechanisms of the human foetus, if we are to be in any position to assess scientifically the teratogenic potential of new chemical substances or other environmental agents. Good epidemiological data are much more difficult to collect for congenital malformations than for mortality, and teratogenicity testing, as we have seen, bristles with problems. At present we do not even understand how most of the known teratogenic agents induce abnormal development in the foetus; I suggest that this is because we still know too little about the mechanisms of normal development in our own species.

Proc. R. Soc. Lond. B. **205**, 111–120 (1979)

Printed in Great Britain

Detection of risk of cancer to man

By R. Peto

*Department of the Regius Professor of Medicine, Radcliffe Infirmary,
Oxford OX2 6HE, U.K.*

Epidemiology can pick out large-scale determinants of human cancer, such as smoking. Also, epidemiology can pick out carcinogens such as asbestos to which groups of perhaps a few hundred or a few thousand workers have been heavily exposed for decades. However, if highly exposed groups cannot be studied then epidemiology cannot recognize carcinogens which, although perhaps widely distributed, produce only a small percentage increase in particular cancers.

Almost all of the environmental pollutants that can affect human cancer incidence will do so only to a very minor extent, at the levels to which we are currently exposed. For this reason, and also because it is often difficult to define an exposed and an unexposed group which do not differ in other ways as well, it will almost always be impossible to do anything epidemiologically except to set a very crude upper limit on their likely hazards. The only way, therefore, to get any direct estimate of these hazards is by laboratory studies of the effects of high doses on various model systems.

For this and for other reasons, it would be highly desirable to have good laboratory models for human carcinogenesis. The characteristics required of satisfactory laboratory systems are reviewed, and it is argued that systematic errors may arise unless one studies epithelial cells from large, long-lived species under conditions of chronic, low-dose exposure to noxious test agents in conjunction with standard chronic doses of agents which may be synergistic with the test agents. (Carcinogenic mutagens may be synergistic with carcinogenic non-mutagens.) For reasons of expense and speed, such studies must be done *in vitro*. If such in-vitro systems can be developed, either by using tissue explants or cell cultures, an important criterion which they will have to satisfy to be trusted will be that under chronic exposure the rate of transformation should be proportional to something like the fourth power of exposure duration.

This paper chiefly reviews the reasons for choosing these specifications for a trustworthy in-vitro model for human carcinogenesis.

Carcinomas and other tumours: carcinomas are what matter most

There is general agreement that most of the cells which cover our skin and which line our gut, ducts and glands should be called 'epithelial cells'. However, there are certain cells which some histologists would prefer to call 'epithelial' while others would not. This is, unfortunately, simply a matter of preferred definition, for there are no generally agreed fundamental principles involved. In other words, the common term 'epithelial' is slightly imprecise.

The term 'carcinoma' is similarly imprecise: there is general agreement in most cases, but there are a few disputed exceptions. Approximately, 'carcinoma' means 'malignant cancer arising from epithelial cells', and this is the definition I shall adopt (thus excluding 'teratocarcinoma' and the frog's 'Lucké adeno-carcinoma'). Whatever definitions are chosen for these two terms, it will be found that while epithelial cells constitute only a small fraction of the human body, most human cancers are carcinomas. Let me therefore divide the cells of the human body into three groups:

1. Epithelia of sex-specific organs (such as breast, penis, cervix uteri, prostate, ovary, endometrium): 20 % of fatal British cancers are carcinomas arising from these epithelia.

2. Epithelia of other organs (such as digestive tract, skin, secretory organs): 70 % of fatal British cancers are carcinomas arising from these epithelia.

3. All non-epithelia (blood, blood vessels, brain, bones, muscles, dermis, etc.): only 10 % of fatal British cancers (including leukaemias, lymphomas, sarcomas, teratomas, melanomas and some other tumour types) are non-carcinomas, i.e. cancers arising from non-epithelial cells.

My first point is that when we talk about 'the causes of cancer', group 1 carcinomas, group 2 carcinomas and group 3 tumours must not be mixed or the whole subject will seem more baffling than it needs to be.

Obviously, because they are the most numerous, group 2 carcinomas are the most important. When we examine them, we find a remarkable uniformity in their age distributions. Suppose we examine a cohort of people who were all born at approximately the same time, to see how their cancer incidence rates vary with age. For most of the group 2 carcinomas, real incidence rates, corrected for data errors (Peto 1977), rise steadily with increasing age, at least up to age 80, being approximately proportional to t^4, t^5 or t^6, where t is age in years. The only major exceptions to this in group 2 are those respiratory sites (e.g. the bronchi) where the numbers of carcinomas are predominantly due to the effects of smoking. These, however, are governed by a power-law relation with 'years of smoking' instead of 'age'. Thus, among regular cigarette smokers who started smoking at age 20, the extra risk of bronchial carcinoma is approximately proportional to $(t-20)^4$, at least up to age 80 (Doll & Peto 1978). Among ex-smokers, the extra annual incidence rate remains approximately constant (probably to within a factor of less than 2) from the time when smoking ceased. Thus, although uniform power-law relations might well govern almost all group 2 carcinomas if the relevant epithelial cells were in a constant environment, age-related changes in this environment can distort this underlying simplicity.

Now, carcinomas usually arise from single epithelial cells; more than one independent alteration is needed *in vivo* to change an epithelial cell into a cancer cell; and, finally, if these several independent changes are all approximately unrelated to age then multi-stage model theory tells us to expect an approximate power-law relation. Given all this, it is reasonable to hope that multi-stage models will prove to be a useful framework in which to make an eventual synthesis of the group 2 carcinoma mechanisms which have been or which will be discovered.

This synthesis may also carry over largely unchanged to the group 1 carcinomas, for just as variation in smoking habits could distort the age-specific incidence rates for lung cancer, so the gross age-related changes which maturation and use impose on the sex-specific organs might be responsible for age-related variations in the rates at which the stages proceed in their epithelia. N. E. Day (personal communication) has produced an excellent fit to breast carcinoma incidence rates by using such hypotheses. Such hypotheses could in principle also account for the shape of the age-specific incidence of carcinoma of the uterine cervix, a disease far commoner in women who have been sexually active than in nuns.†

In summary, although data on the group 1 (sex-specific) carcinomas may be complicated by age-related changes in the structure and usage of the sex-specific organs, multi-stage models similar to those applicable to group 2 (all other) carcinomas (but with age-related event rates) may again provide a useful framework for an eventual synthesis of the processes involved in their genesis.

However, the age distributions of the several group 3 tumour types (i.e. the non-carcinomas) are so heterogeneous that I doubt whether it is wise to attempt any unified description of their mechanisms of induction. Moreover, simple multi-stage models (with no intrinsic relevance of age *per se*; see Peto *et al.* 1975) predict approximate power-law relations between cancer incidence rates and age, but this is not found for most group 3 tumour types. While, therefore, it is reasonable to hope that a unified synthesis of the mechanisms of production of 90 % of human cancer (the carcinomas) will eventually be achieved, probably by using the general framework of a multi-stage model to put the parts of it together, this description may not carry over to some of the remaining 10 % (the non-carcinomas). Conversely, laboratory phenomena which have only been demonstrated in non-carcinomas may have little relevance to *malignant* cancers of *epithelial* cells. By this criterion, sarcoma viruses, leukaemia viruses, lymphoma viruses, papilloma viruses, tumour dedifferentiation and terminal differentiation may be irrelevant to 90 % of human cancer.

† For example, suppose that, on average, British women make love from the age of about 20 to the age of about 50 and suppose that partially transformed epithelial cells in the uterine cervix have a selective advantage over their unaltered neighbours for only as long as regular sexual intercourse continues. One might then expect (Peto 1977) incidence rates of cervical cancer to rise rapidly between age 20 and age 50 and then to flatten off, just as is actually observed. (It may enhance the plausibility of such post-hoc 'explanations' of the age-specific incidence rates of group 1 carcinomas to note that cigarette smokers who smoke one pack a day from the age of 20 to the age of 50 and who then give up have also been observed to suffer extra lung cancer incidence rates which increase in proportion to $(t-20)^4$ between age 20 and age 50 and which then become fairly flat after age 50.)

Acute high and chronic low exposure: chronic low exposure matters most

Experimentally, it is easy to give a single acute dose of a carcinogen which suffices to give cancer to a fair proportion of the animals exposed to it. Likewise, acute exposure of a human organ to a few hundred rads† may double the probability that cancer will arise in it. However, for most of the environmental agents which concern us single acute exposures producing significant risk will be uncommon: chronic or recurrent exposure to low levels of risk are a more likely hazard. I wish to suggest that if we want to understand the effects of such chronic low exposure on humans then study of the effects of acute high exposure is a really bad model, because in almost every situation which has been considered in detail acute high exposures also have serious side-effects on the target organ other than the direct carcinogenic effect which is to be studied. For example, Potten (1977) has demonstrated that in the rodent intestine there is a population of cells that may well be the real targets for carcinoma induction, 50 % of which are killed by 20 rad. Likewise, the single doses of initiator commonly used in two-stage mouse skin carcinogenesis often suffice to cause severe ulceration, removing the basal layer of the epithelium entirely and requiring regrowth (which itself has a promotional effect). Another example is intragastric instillation of some milligrams of DMBA to produce mammary tumours in rats; such doses seriously damage subsequent hormone secretion, which itself affects rat mammary tumorigenesis.

Usually, I suspect, when environmental contaminants affect human cancer, they do so by chronic exposure to dose rates so low that the structure of the target organ is not greatly upset. There are some obvious exceptions to this (e.g. smokers' lungs and Iranian oesophaguses), but as a general principle I believe that our first task should be to elucidate the effects of chronic low exposure to carcinogens, and that one should not be unduly disturbed if one is unable to explain effects (e.g. curious dose–response relations) that are only reported from experiments with acute insults, or very strong insults.

Time and lifespan: large, long-lived species may have fundamentally different epithelia

Having decided that we want to understand *carcinoma* induction by *chronic* exposure of epithelial cells, in what system should this be studied? The choices a scientist must make are: (*a*) *in vivo* or *in vitro*? and (*b*) in humans or in some other species?

The natural answer for guaranteed relevance is *in vivo*, in humans. This, of course, is the epidemiologist's answer, and so far, in cancer as in other diseases, epidemiologists have had much greater practical success, and at much lower cost,

† 1 rad = 10^{-2} Gy = 10^{-2} J kg^{-1}.

than all the fundamental scientists together have had. Simple cigarette smoking accounts for one-third of all British cancer, and would account for more were it not for the expedient, devised in response to epidemiology, of making cigarette paper perforated or porous so that noxious smoke is diluted by air. The international correlations of fat consumption with breast and colon cancer (which would not have been suggested by any current in-vitro studies) will probably lead to findings which push the proportion of cancers for which *preventable* causes have been identified by epidemiology to well above 50 %. Thus, despite being unable to experiment deliberately (except in randomized trials of prevention), the epidemiological study of human tumours *in vivo* will lead to strategies which may win more than half of the battle against cancer. The successes and difficulties of epidemiology are documented in much more detail in Doll (1977, 1978), articles which should be widely read; I do not wish to review them.

Curiously, despite this praise for epidemiology, my main purpose is to consider (as a complete outsider) how laboratory research might best help towards preventing chronic environmental pollution from causing human carcinomas. Let me return to the two questions of *in vivo* or *in vitro*? and in which species?, eschewing epidemiology. Two very important considerations that are usually overlooked in selecting experimental systems are size and lifespan. A living human has perhaps 10^{12} viable epithelial cells, and a mouse perhaps 10^9. Although the human lifespan is perhaps 75 years, compared with perhaps $2\frac{1}{2}$ years for a mouse, the ultimate risk of carcinoma in both mice and men is of approximate order 10^{-1}. Now consider a single epithelial cell aged t years. If it is a mouse epithelial cell, then the probability that it will generate a carcinoma tomorrow is, of course, proportional to something like t^4, where t is the mouse's age, the constant of proportionality being of approximate order 10^{-14}. If it is a human epithelial cell, however, then this constant of proportionality is more than 10^9 times smaller. Now, a factor of 10^9 is so extreme as to suggest that mechanisms may be at work in human epithelial cells that are qualitatively different from those in mouse epithelia. This factor of 10^9 is the product of the 1000-fold ratio of the two body masses and about the fifth power of the 30-fold ratio of the two lifespans, but it is not strongly dependent on the observed ratio of eventual carcinoma risk in mouse and man. Such calculations suggest that the *in vivo/in vitro* distinction may often be less important than the overriding need to use as experimental models epithelial cells from species with a lifespan and body mass comparable to that of man.

Although this line of argument casts doubt on the relevance of almost all studies *in vivo* (since only large, long-lived mammals are allowable), I believe it to be sound. (It may underlie the observation that transformation *in vitro* of rodent epithelial cells is fairly easy, while transformation *in vitro* of human epithelial cells is almost impossible.) If accepted, it suggests that there should perhaps be more effort to solve the difficult technical problems of how to establish human (or primate) epithelial cells with a stable karyotype *in vitro* (as Sun & Green 1977 have been doing) and how to transform them by steady chronic exposure to low

doses of various insults to give a power-law relation between cellular transformation and time. If this could be achieved, it might well be a most valuable
experimental model in which to study the modifying effects of various environmental contaminants, and its establishment might lead to the recognition of new
classes of carcinogens that are not mutagens.

CLASSES OF CARCINOGENS: SOME ARE NOT
BACTERIAL MUTAGENS

It has been argued that carcinogens are mutagens, but there is fairly good
evidence that for carcinomas this is not always true. Perhaps mutagens are usually
carcinogens (if they get a chance to act on the epithelial stem cells) but even if
they are, there is almost certainly another class of rate-determining causes of
human and animal carcinomas which are not, even after suitable metabolic
activation, bacterial mutagens and which do not act by activating or transporting
such mutagens. I have reviewed some evidence for this statement (Peto 1977)
where details and references may be found. The statement may be illustrated by
comparing the effects of smoking two packs of cigarettes daily for 20 years with
the effects of smoking one pack daily for 40 years. The latter is a considerably
more hazardous activity; in other words, the cancer risk depends much more
strongly on duration of exposure than on daily dose-rate. The same is found for
experimental animals: after administering a chronic dose of x units of carcinogen
per day for y years the incidence rate of carcinomas is typically proportional to
$x^a y^b$ where a is about 1 or 2, while b is of about 3 or, usually, more. Genetic hetero-
geneity in the population being studied can bias the observed values of both b
and a downwards, but Sarah Parish (personal communication) has shown that it
has little effect on their ratio. The fact that b is nearly always at least twice a in
both animal and human data suggests that this would also be true in a population
of genetically identical individuals. Why should risk depend so much more strongly
on time than on the daily dose rate of the test carcinogen? The most likely explanation, I would suggest, is that there are processes involved in carcinoma induction which proceed largely independently of the concentration of mutagen, and
that longer duration of exposure give these 'spontaneous' changes more time to
accumulate. (For example, if mutagens cause recessive changes in a diploid cell,
then rearrangements of the diploid state may be necessary before the mutation
can affect the behaviour of the cell, and agents which, although not mutagenic,
stimulate such rearrangements might then increase carcinoma incidence rates.)

All that I wish to argue is that there are classes of agents, other than mutagens
and their metabolic precursors, with the ability to affect human carcinoma incidence rates multiplicatively, and that to identify them we should try to study
laboratory models which do not differ in some profound way (epithelial/non-
epithelial; chronic/acute; large and long-lived/small and short-lived) from the
system that we wish to model. Having identified and understood a class of such

agents by the study of realistic systems, we *may* then discover how to assay future chemicals for activity as a member of that particular class by some simpler and more direct test system, but while, as now, whole classes of agents seem to be eluding us we must try to make our in-vitro systems more realistic.

USE OF IN-VITRO TESTS: LIMITATIONS OF EPIDEMIOLOGY

As Doll (this symposium) has demonstrated by his review of British trends in mortality, whatever effects environmental pollution may be having on mortality in general and cancer in particular, they are not at present likely to be gross, and the gross international differences in cancer which currently exist are unlikely to be attributable to environmental pollution in any sense in which this is ordinarily understood. (Some of the numerically large international differences are probably due to tobacco, alcohol, sex, excess dietary intake of certain components, shortages of certain others, reproductive history and sunlight, but none can be confidently attributed to DDT, saccharin, or air or water pollution.) I do *not* wish to argue that environmental pollution is irrelevant, merely that the effects (if any) of most pollutants on specific diseases are likely to be rather minor (an increase of a small percentage, for example, rather than a tenfold increase). This means, unfortunately, that epidemiology will probably be impracticable unless highly exposed groups (e.g. factory workers) can be studied, for while epidemiology can usually monitor twofold relative risks (if 'exposed' and 'unexposed' groups can be clearly identified), it can rarely monitor a 10 % increase in hazard reliably, not only because sample sizes would have to be very large but also because the intrinsic biases of the non-experimental method are of at least this order of magnitude. Therefore such minor effects can only be monitored by laboratory studies, and for the reasons already cited I fear that current in-vitro testing methods may systematically miss important classes of human carcinogens. (Saccharin, perhaps, might show up clearly if it were tested to see whether it was synergistic with mutagens.)

I do not know whether the story that in about 1950 peanut butter manu-facturers considered using dimethylnitrosamine as an emulsifier at one stage in the process is true, but if it is then laboratory research might have already scored a huge success in preventive medicine, unnoticed by almost everybody. A more typical story where laboratory studies have probably prevented some deaths from a cause that would not have been recognized for decades by epidemiologists con-cerns Tris. This is a flameproofing material, added 20 % by mass from 1972 to 1977 to most American children's pyjamas. It was found to be mutagenic to bacteria, carcinogenic to rodents and absorbable through human skin. Other flame-retardants exist, and it is likely that the recent legislative removal of Tris from American clothes will save hundreds or thousands (but probably not hun-dreds of thousands) of lives. This seems to me to be a model of what we may reasonably hope for from laboratory tests, and it is one reason why I should like to see more realistic in-vitro models. The other reason, of course, is that my real interest is in how the common cancers arise, and if we could discover a new class of carcinogens which are synergistic with the mutagens (e.g. by studying the effect of adding various concentrations of the test substance on human epithelial cultures bathed in a low but constant concentration of a standard mutagen) then this would be scientifically very interesting, and might be of substantial practical value if it gave epidemiologists useful clues as to what to examine most closely when studying the common cancers.

FURTHER READING

The main reference, which gives details to support all the arguments I advance, is Peto (1977). Other relevant references are Sir Richard Doll's paper in this Symposium, which demonstrates that British national trends in mortality are generally reassuring; Doll (1977), which reviews the epidemiological findings about cancer; Peto *et al.* (1975), which demonstrates experimentally that power-law relations between carcinoma incidence and age can arise in the absence of any greater susceptibility of old animals to the action of carcinogens; and Cairns (1975), which is simply interesting.

REFERENCES (Peto)

Cairns, J. 1975 Mutation selection and the natural history of cancer. *Nature, Lond.* **255**, 197–200.

Doll, R. 1977 Strategy for detection of cancer hazards to man. *Nature, Lond.* **265**, 589–596.

Doll, R. 1978 An epidemiological perspective of the biology of cancer. *Cancer Res.* **38**, 3573–3583.

Doll, R. & Peto, R. 1978 Cigarette smoking and bronchial carcinoma: dose and time relationships among regular smokers and lifelong nonsmokers. *J. Epidemiol. Community Hlth* **32**, 303–313.

Peto, R., Roe, F. J. C., Lee, P. N., Levy, L. & Clack, J. 1975 Cancer and ageing in mice and men. *Br. J. Cancer* **32**, 422–426.

Peto, R. 1977 Epidemiology, multistage models and short-term mutagenicity tests. In *Origins of human cancer*, pp. 1403–1428. New York: Cold Spring Harbour Publications.

Potten, C. 1977 Extreme sensitivity of some intestinal crypt cells to X and γ irradiation. *Nature, Lond.* **269**, 518–521.

Sun, T.-T. & Green, H. 1976 Differentiation of the epidermal keratinocyte in cell culture. *Cell* **9**, 511–521.

Sun, T.-T. & Green, H. 1977 Cultured epithelial cells of cornea, conjunctiva and skin: absence of marked intrinsic divergence of their differentiated rates. *Nature, Lond.* **269**, 489–493.

Discussion

B. A. BRIDGES (*M.R.C. Cell Mutation Unit, University of Sussex, U.K.*). One of the requirements proposed for a valid in-vitro test system is that the rate of transformation should be proportional to approximately the fourth power of exposure duration. What grounds are there for believing that this type of response is solely a property of the cells and that it is not, at least in part, determined by the response of the whole animal towards malignant or premalignant cells?

Secondly, by choosing cells from a long-lived animal would one not expect, if the latent period were solely a property of the cell and not of the cell–host system, that transformation *in vitro* would take years, in which case the value of the system as a short-term test would be arguable? While Mr Peto's criteria may be fine if one wishes to approach the process of human cancer induction *in vitro*, I suggest that rather different criteria should apply if one wishes to devise a better test for potential human carcinogens, not least of which are practicability and a more realistic imitation of human detoxification and activation pathways.

R. Peto. First, the 'fourth power' criterion was suggested precisely as a check that important rate-determining effects dependent on the whole animals (e.g. ability of transformed cells to metastasize, or to elicit angiogenesis) are *not* being missed. There are, however, good grounds for believing that the power-law relation does not owe its existence to any age-related changes in the susceptibility of the whole host, and these are presented in Peto *et al.* (1975).

Secondly, my point was not that the in-vitro systems which I propose would be the best short-term tests for any one particular class of agents, but rather that the study of such systems could well lead us to identify a new class of non-mutagenic carcinogen. If this happens, then we *may* soon realize that we can assay new chemicals for membership of this class in some much more direct way. However, until we know the class of agents that we are looking for we cannot devise direct 'practicable' tests for membership of that class.

While I agree that it is highly desirable that realistic human activation pathways (over and above those already present in the epithelial cell itself) should be included, I do not see any obvious ways in which this can be done reliably until we know the class of agent we are seeking. Also, I do not agree that realistic detoxification pathways need to be included; indeed, at our present state of knowledge they might even be undesirable.

Finally, on the question of 'practicability', I am uncomfortably aware that the multi-stage model arguments that I have used are only theoretical and may therefore be seriously misleading. However, when theories are suggested the scientific community can make one or other of two opposite scientific errors. The more obvious error, and the one to which mathematicians who dabble in science are most prone, is being happy to develop the detailed consequences of abstract theories without demanding experimental verification. The less obvious error, which at present seems to me to be the greater danger in cancer research, is that experimental biologists may be so little attracted by theoretical speculations (which of necessity will oversimplify reality) that they do not follow up the experimental leads that are suggested by them.

By instinct, many biologists do not feel that the ill-defined 'stages' (changes with little phenotypic effect in a few scattered stem cells) discussed in multi-stage models are real biological phenomena which it would be 'practicable' to study. Because of this, I have tried to suggest questions that might yield interesting answers even if my reasons for asking them are largely mistaken: (1) Why is there a difference of a factor of 10^9 between carcinoma induction in mice and men? and, (2) What agents (and, eventually, what classes of agents) act synergistically with Ames-type mutagens in carcinoma induction? The pursuit of in-vitro human epithelial cell cultures that I have suggested, to be followed eventually by an attempt to reproduce in them some aspects of the kinetics of in-vivo carcinoma induction, may or may not be the most practicable experimental approach to these questions (although the growth *in vitro* of human epithelial cells is a worthwhile scientific objective anyway). No approach can be guaranteed to be practicable until it has been tried, and if we restrict ourselves to approaches that we already know to be practicable then the practicable may be the enemy of the possible.

Proc. R. Soc. B. **205**, 121–134 (1979)

Printed in Great Britain

Carcinogen prediction in the laboratory: a personal view

By R. C. Garner

Cancer Research Unit, University of York,
Heslington, York YO1 5DD, U.K.

Although carcinogens can be divided into various categories, i.e. viruses, physical agents and synthetic and naturally occurring chemicals, it is the latter that give rise to the greatest concern because of their number, quantity and distribution. Present methods of testing chemicals for potential carcinogenicity rely in the main on administration of these at maximally tolerated dose levels to animals, usually rodents, for the animals' lifetime. Such tests would be economically impractical for all chemicals to which man is exposed. New methods have recently been introduced to screen large numbers of chemicals quickly and cheaply which rely on the unifying hypothesis that all carcinogenic chemicals are electrophiles or must be converted to such by metabolism. These methods will be reviewed and compared with traditional methods of carcinogenicity testing, particularly as to their role in attempting to predict hazard to man.

Varying cancer incidence rates in different parts of the world, in different areas of countries and even between closely situated towns suggest that much human cancer may be largely determined by external factors. Individuals could respond in differing ways to these factors, thus accounting for the variation in cancer incidence within the same population. Similarly, studies of cancer incidence in immigrant populations show that individuals tend to lose the cancer site-spectrum and incidence characteristic of the country that they have migrated from and assume that of their new home. These facts suggest that cancer is to a large extent environmentally determined and therefore preventable (Doll 1967; Doll & Vodopija 1973; Saffiotti & Wagoner 1976; Doll 1977). This premise is increasingly being accepted for a number of reasons including: (1) cancer families are relatively rare; (2) there is little evidence of infectious transmission of human cancer; and (3) particular types of cancer can be explained by exposure to specific agents.

If one accepts the above statements, one can then go on to discuss how 'cancer' might be prevented. I wish in this paper to describe present methods of carcinogenicity testing which enable us to identify specific chemicals that can initiate cancer as well as methods that might be used in future screening programmes. It should be borne in mind at the outset that the rigour of any testing protocol will depend entirely on the exposure level to particular chemicals, on the duration of exposure and the number of people exposed to any particular substance(s).

Testing procedures for a food additive or pharmaceutical have in the past been more extensive than for industrial chemicals because of the greater numbers of people exposed and the limited numbers of chemicals involved. This policy is now accepted as being socially unacceptable because it has meant that many industrial chemicals to which people were exposed had never been tested for toxicity. Nevertheless it is a sobering thought that if industrial chemicals were subjected to the same rigorous (perhaps too rigorous) toxicity tests as those required for pharmaceuticals, the cost in the United Kingdom alone would be some £25 × 10^9 to test the backlog of chemicals and some £250 × 10^6 p.a. thereafter. (These figures are based on a cost of £500000 for a complete toxicity study; costs of carcinogenicity assays constitute approximately one-fifth of these figures.) Whatever one's views about the desirability or otherwise of conducting particular tests, it should be remembered that it is the consumer who pays, so that the money involved becomes an overriding factor. Profit margins are so small on some industrial chemicals that the cost of a pharmaceutical type toxicity test would not be commercially sensible and so the manufacture of the chemical would be discontinued with the consequent social effects. One cannot condone the manufacture of hazardous chemicals where there is human exposure but an attempt must be made to decide what is acceptable and what is not. Such decisions will have to be taken not only by scientists but by economists and others who are concerned with risk–benefit analysis. Clearly we cannot muddle along in the future in the same way that some companies do now or have done in the past in terms of determining the biological effects of chemicals. It is simply not acceptable to carry out a deliberate human toxicity study.

ANIMAL CARCINOGENICITY STUDIES

What evidence is there to support the concept that cancer in man is in any way predictable? The strongest evidence must come from the finding that of those compounds or treatments that have been identified as causing human cancer, all with the possible exception of arsenic have been shown to be carcinogens in one animal species or another (see table 1). The question which is of prime importance is: are there any animal carcinogens which are not human carcinogens

TABLE 1. HUMAN CARCINOGENS ACTIVE IN ANIMALS

asbestos	bischloromethyl ether
ionizing radiation	benzene
ultraviolet light	mustard gas
2-naphthylamine	vinyl chloride
4-aminobiphenyl	stilboestrol
benzidine	aflatoxins
chrome ore	cyclophosphamide
nickel ore	synthetic steroids
polycyclic hydrocarbons	

and vice versa? It is on this point that there is a great deal of heated discussion at present over compounds such as amaranth, saccharin, cyclamate, phenobarbitone and so on. Of these compounds only the latter appears to have been tested adequately (Ponomarkov *et al.* 1976; Rossi *et al.* 1977), data on the others have often been of a preliminary nature. In the final analysis, only epidemiological studies will tell us if there is any risk associated with these chemicals; no amount of animal experimentation can do this.

There have been many reviews, meetings and conferences on the extrapolation of animal carcinogenicity data to man. What emerges is that there are only a few compounds that have been adequately tested for carcinogenicity (see I.A.R.C. *Monographs on evaluation of carcinogenic risk of chemicals to man*). These animal studies must, however, provide the data base to validate all of the newer short-term tests for carcinogenicity that I shall discuss.

I do not wish to describe long-term animal studies in any great detail since I am not an expert on them. It is enough to say that the results obtained can be dependent on, for example, the animal species used (Clayson & Garner 1976), its sex, strain (Rueber 1976), nutritional status (Wattenberg *et al.* 1976), hormonal balance, purity of the compound tested, and numbers in the test and control groups (Fears *et al.* 1977). Despite this, the long-term feeding of chemicals to animals is our only well studied experimental method of determining the carcinogenicity or otherwise of a chemical. If one suspects that a particular compound is a carcinogen in humans the most definitive test to establish this must involve animal exposure. Such tests, as far as one can see, will detect all classes of agent known to be carcinogenic, i.e. physical, hormonal, viral and chemical, in contrast to some other of the tests which I shall describe.

ARE THERE ANY UNIFYING MECHANISMS OF CARCINOGENESIS?

Of the compounds listed in table 1 most are organic chemicals. I wish in the main to look at carcinogens in this class, primarily because their mechanism of tumour initiation is thought to be understood. That is, the initial interaction(s) of the organic chemical with biological material within the body is known. This is a long way from saying that we understand the mechanism of tumour production, but we can say at present what the first step is likely to be in the cancer process. This first step may also allow us to predict, on the basis of chemical structure, whether a compound is a carcinogen or not.

Administration of carcinogens to animals results often in the covalent binding of the chemical to macromolecules within the body. On the whole, the highest level of binding is found in organs susceptible to the particular carcinogen. These covalently bound adducts arise through metabolism of the carcinogen to a re-active chemical species (an electrophile) and subsequent reaction with nucleo-philic sites contained within macromolecules. Figure 1 shows how aflatoxin B_1, a human carcinogen, is activated by liver mixed function oxidase enzymes to a reactive epoxide and also shows the structure of the major adduct found in nucleic acid consequent upon its reaction with guanine (Martin & Garner 1977). Carcinogenic susceptibility to aflatoxin B_1 could be dependent on how much epoxide metabolite is produced and how much reacts with critical target macromolecules. This concept, of conversion of relatively inert chemicals to reactive species, has been proposed by several authors, particularly the Millers at the University of Wisconsin (Miller 1970), to be a common feature for most if not all chemical carcinogens (figure 2). Certainly the classes of chemical carcinogen which appear to fulfil this criteria are impressive. One can include the polycyclic

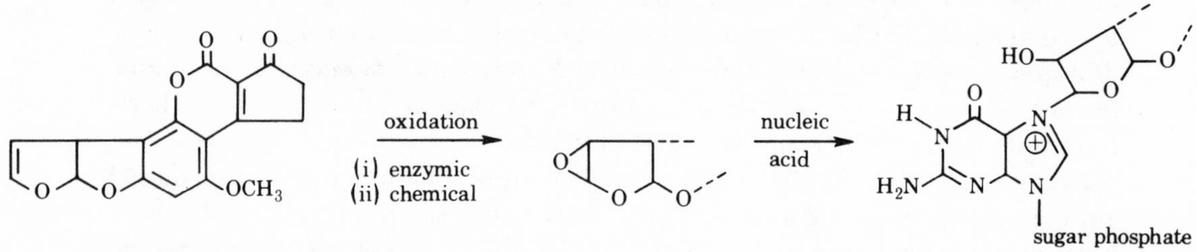

aflatoxin B_1 8, 9-dihydro-8-(N^7-guanyl)-9-hydroxyaflatoxin B_1

FIGURE 1. Activation of aflatoxin B_1 and subsequent reaction with nucleic acid.

aromatic hydrocarbons (Brookes & Lawley 1964; Sims & Grover 1974), various fused or conjugated aromatic amines (Clayson & Garner 1976), the nitrosamines (Magee et al. 1976), certain mycotoxins and natural products and a number of heterocyclic compounds. What is also impressive is that all of these classes react extensively with nucleic acids as well as proteins. Nucleic acid reaction is considered at present to be more important than protein reaction particularly because of the lack of experimental evidence to suggest that reaction with proteins can account for the heritability of cancer in daughter cells. A further reinforcement of the importance of nucleic acid reaction is the finding that the most potent animal carcinogen, aflatoxin B_1, is bound much more extensively to nucleic acid than to protein and that in a resistant species, the hamster, amounts of liver protein-bound carcinogen are similar to those in the rat, a susceptible species (Garner & Wright 1975). Naturally, gross differences in amounts of carcinogen binding can only give a crude estimate of the importance of reaction with any macromolecular type. What is also striking is the relation between chemicals that are carcinogenic in animals and those mutagenic in bacteria, indicating that nucleic acid reaction could well be a prerequisite for tumour production.

Electrophilic metabolite generation from carcinogens can therefore be presumed to be essential for cancer initiation for many organic chemicals. The events which determine whether an initiated or pre-neoplastic cell progresses to a frank malignancy are largely unknown. My discussion will therefore centre around what is known, namely that electrophilic metabolites initiate cancer, rather than the unknown events which determine whether a pre-neoplastic cell progresses to a fully malignant cell.

METHODS OF DETECTING ELECTROPHILIC METABOLITES *IN VITRO*

Since electrophilic metabolites of carcinogens react readily with nucleophiles, measurement of such reactions gives a means of monitoring electrophilic metabolite production. Figure 2 shows various biological methods of detecting carcinogen

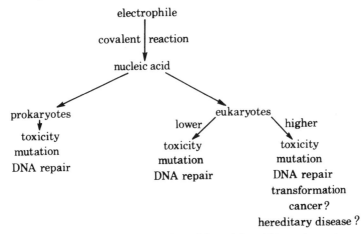

FIGURE 2. Consequence of electrophile reaction with nucleic acid in prokaryotes and eukaryotes.

reaction with nucleic acid in both prokaryotic and eukaryotic cells. The various techniques to be described have been reviewed by a number of authors (Magee 1974; Stoltz *et al.* 1974; Bridges 1976).

(a) Nucleic acid reaction in prokaryotes

Mutation as a result of exposure to an alkylating agent has been known for 30 years, but it is only in the last 6 years that the relation between mutagens and carcinogens has been proposed (McCann *et al.* 1975). This is because there was for a considerable time a failure to appreciate that carcinogens require metabolism to exert their effects and that most prokaryotes are unable to carry out these steps. Once it was accepted that oxidative metabolism was essential, this area expanded dramatically, so that at present there are few human or animal carcinogens which have not been shown to be mutagenic. It is perhaps natural that with this impressive correlation between mutagens and carcinogens there is now a tendency to assume that a particular substance is a hazard merely on the basis of its ability to induce mutations in bacteria. This is extremely unfortunate, because there is likely to be a strong reaction against this idea, the pendulum then swinging in the other direction so that it will be extremely difficult to convince people that any mutagen is a carcinogen.

Of the tests with prokaryotes in current use, mutation to prototrophy in *Salmonella typhimurium* is probably the most intensively investigated at the present time. The assay with the use of this microorganism involves co-incubation of a rat liver preparation with the carcinogen in the presence of the bacterial tester strain, the ingredients being contained within a soft agar overlay (McCann *et al.* 1975) on a minimal agar plate or in liquid suspension (Garner *et al.* 1971; Malling 1971). The bacterial strains are all histidine auxotrophs, mutation being scored by counting the number of revertant colonies. A series of tester strains has been of theoretical objections can be made to this procedure but the fact is that at present constructed, allowing different classes of carcinogen to be detected. A large number an impressive correlation is found between compounds that induce mutation and their animal carcinogenicity. Of the *Salmonella* tester strains in use, I tend to view results obtained with strains TA1537 and TA98 with more caution than those obtained with TA1535 and TA100. This is because the latter two strains are base-substitution mutants. Reversion usually arises in these strains though covalent reaction of a carcinogen with nucleic acid, so disrupting normal base pairing or through error-prone repair pathways. The former two strains are both frame-shift mutants, and are reverted by intercalating agents such as the acridines. There is no evidence that these compounds are carcinogens until an alkylating moiety is attached to them. Since the hypothesis of metabolic activation states that most carcinogens are electrophiles, one has to be sure that the mutation observed is as a result of covalent DNA binding (Garner & Nutman 1977).

Other bacterial strains used for mutagenicity testing are tryptophan auxotrophs of *Escherichia coli* with varying DNA repair activity, and a multiple mutant strain of the same organism in which both forward and back mutation can be scored (Green & Muriel 1976; Mohn & Ellenberger 1973).

A simpler approach to detecting DNA damage in prokaryotes is to measure survival of isogenic bacteria with varying DNA repair capacities. Convenient tester strains are available for this type of assay, which depends on the fact that survival after exposure to DNA damaging agents will be dependent on the DNA repair capacity of the particular strain. The most sensitive tester strain to such agents would be one which is lacking both the *uvr* endonuclease and is also re-combination deficient (Bridges *et al.* 1972). A comparison of the sensitivity of this strain with strains lacking only one of these repair pathways or the wild-type strain would show whether or not toxicity was due to DNA damage. Assays of this type are the so-called *rec-* or *pol-*assays (Kada *et al.* 1972; Slater *et al.* 1971). Neither of these have been extensively validated, although both appear to be suitable only for direct acting alkylating agents rather than those requiring meta-bolic activation. By using a liquid suspension assay and *E. coli* repair deficient strains, it was possible to show that the toxicity of an aflatoxin B_1 metabolite, the production of which was mediated by rat liver, was due to DNA damage rather than increased penetration of the activated metabolite (Garner & Wright 1973).

Finally, it is possible to measure DNA repair synthesis in bacteria as a result of carcinogen damage. This might have two advantages over mutagenicity assays in that only a single tester organism need be used and increased DNA repair could only arise through covalent reaction of the agent with nucleic acid (Thielmann 1976).

(b) *Nucleic acid reaction in eukaryotes*

Many of the assays of eukaryotes are basically the same as those that I have described for prokaryotes. What distinguishes these assays from bacterial assays is their increased complexity. Whereas a bacterial assay takes only 48 h to carry out from start to finish, a mammalian cell assay can take up to 6 months, depending on type.

There is now little doubt that mutations can be induced in mammalian cells in culture. Of the various techniques in current use, mutation to resistance to nucleic acid analogues appears to be the most widely preferred method. Resistance can be to compounds such as 8-azaguanine and 6-thioguanine (DeMars 1974; Shin 1974; Nikaido & Fox 1976). Recently, mutation to an alanine requirement has been demonstrated as well as reversion of the mutant cells by ultraviolet light, gamma rays and 4-nitro-quinoline-*N*-oxide (Suzuki & Okada 1976). For all of these mutation assays it is necessary to show that the mutant cells scored have a heritable mutation and that resistant cells do not arise through some other epigenetic mechanism.

Although only a few carcinogens have shown mutagenic activity in mammalian cells, this may be due to cells in culture having a low level of metabolizing enzyme activity. This problem can be partly overcome by the use of feeder cells but only if these have the necessary enzyme activity. To date, mutation in mammalian cells has only been studied extensively with the polycyclic aromatic hydrocarbon group of carcinogens (Huberman & Sachs 1974).

Survival studies with the use of human cell lines deficient in DNA repair have been suggested for carcinogen screening. At present, reports of this method have centred around compounds which act directly rather than requiring metabolism. The use of Xeroderma pigmentosum cells (an inherited human disease characterized by extreme sensitivity to ultraviolet light and a much increased skin cancer incidence) has shown these cells to be deficient in their ability to remove not only ultraviolet light damage (Cleaver 1970) but also chemical damage (Stich & San 1973). Agents which have greater toxicity in Xeroderma cells than in normal cells in culture are thought to be toxic by virtue of their reaction with DNA. The assay is similar therefore to the bacterial *rec-* and *pol-*assays previously described.

Reaction of an electrophile with nucleic acid in mammalian cells results in subsequent DNA repair. This can be monitored (1) by the non-S-phase uptake of [^{3}H]thymidine into nucleic acid measured by auto-radiography (San & Stich 1975) or scintillation counting of the extracted DNA (Martin *et al.* 1977), or (2) by looking for the presence of single strand breaks in the nucleic acid on alkaline sucrose density gradient centrifugation. A recent study has recently been completed at York on the testing of some 50 compounds of known carcinogenic activity for their ability to induce 'unscheduled DNA synthesis' in HeLa cells in culture. No known carcinogen was negative in this study with the exception of safrole. Four compounds, namely urethane, dimethylaminoazobenzene, and dimethylnitrosamine and diethylnitrosamine, which are weakly active or inactive in standard bacterial mutagenicity assays, were active, and two compounds, 9-aminoacridine and sodium azide, which are mutagenic but not carcinogenic, were inactive. This assay would appear to be suitable for inclusion in any short-term testing protocol for carcinogens, providing a complementary assay to bacterial mutagenicity. HeLa cells have a low but perceptible capacity to activate carcinogens since all of those compounds which were active after liver metabolism had a low but significant effect in inducing 'unscheduled DNA synthesis' in the absence of liver. Increased sensitivity might be achieved by induction of the mixed function oxidase enzymes in these cells.

Finally in this section I shall discuss briefly transformation of cells in culture, that is the conversion of cells which display contact inhibition of growth to cells that have lost this property (Berwald & Sachs 1963; DiPaolo et al. 1971; Heidelberger 1973). Injection of the transformed fibroblasts into syngeneic animals should give rise to fibrosarcomas if the cells have genuinely been transformed whereas the original cell lines does not. Just what a transformed cell is, however, is not clear. Nearly all studies have used rodent cell lines which can be relatively easily transformed by chemicals; cultured human cell lines are more difficult if not impossible to transform by chemicals (Ponten 1976). Criteria for transformation are much discussed. Some people think that morphological criteria are sufficient to indicate transformation, others that the ability to grow in soft agar is a reasonable criterion for transformation whereas the purists think that transformation can only be demonstrated by malignant growth of transformed cells in animals.

Further assay procedures for transformation are the use of viruses in conjunction with chemicals. The virus itself does not transform the cell, and neither does the carcinogenic chemical, but the two in combination do (Freeman et al. 1973). Although this technique appears attractive, insufficient data are available at present to determine whether this type of test can be used routinely.

OTHER SHORT-TERM TESTS FOR CARCINOGENS

A great variety of other tests have been suggested to screen carcinogens quickly. Most of these have not been tested with large numbers of compounds. Although certain of them appear to be useful for particular classes of carcinogen, for example sebaceous gland suppression or non-specific esterase activity in sebaceous glands of mouse skin for the polycyclic hydrocarbon group of compounds (Healey et al. 1970), there is no evidence to suggest that they will become widely adopted in any routine testing protocol.

IN-VIVO TESTS OF SHORT DURATION FOR CARCINOGENICITY

Many of the short-term tests described have a major disadvantage in that they use liver enzyme preparations, usually from a rat, to convert the test compound to a reactive metabolite. If activation of a particular compound goes through some mechanism other than mixed function oxidase attack or requires a further metabolic step, such as esterification, after oxidation, the requisite enzymes may not be contained within the liver preparation used. Furthermore, there is much evidence available to support the idea that tests *in vitro* give a measure of carcinogenic potential for a particular compound but not of species sensitivity or resistance. Thus a hamster liver preparation can activate aflatoxin B_1 to its 8,9-oxide to a much greater extent *in vitro* than can a preparation from the rat and yet this former species is resistant to the carcinogenic action of the compound whereas the rat is sensitive (Garner *et al.* 1972). In other words, there is an inverse correlation between carcinogenic susceptibility and the ability to activate this mycotoxin. Why this should be is not clear. It is possible that pharmacokinetic and pharmacodynamic parameters determine aflatoxin B_1 susceptibility and these cannot be mirrored in an artificial in-vitro assay. It is therefore essential, if we are interested in carcinogenic potency and hazard to man, to devise assays which can give us some idea of these two parameters.

What is clear at present is that there is no one animal species which we can say will behave in the same way as man. It has become popular to use non-human primates for this purpose but there is little if any evidence that this will be a useful approach. This brings us full circle therefore to in-vivo studies as the best method of attempting to predict hazard to man. What techniques are available, to speed up the process of carcinogenicity testing in animals, that do not involve the vast expense of traditional methods? Approaches to this problem which have been suggested are the use of strain A mice and the scoring of lung adenomas (Shimkin *et al.* 1966). This approach, while quicker than conventional studies, may be unsatisfactory because of variable spontaneous tumour incidence and the great number of factors which can alter this incidence in the control population.

Another approach is that of scoring the numbers of transformed cells taken from hamster foetuses after administration of the carcinogen to the mother. This assay, which is an in-vivo/in-vitro assay has the advantage of using whole animals but may be subject to problems associated with the species used (DiPaolo *et al.* 1972). Should one use hamsters or rats or mice and what criterion of transformation should be used? A further approach is to treat the whole animal with the test compound and then remove and culture particular organs. In this way the conversion of a pre-neoplastic cell to a frank malignancy can be speeded up (Mondal 1975). This approach seems promising on paper but, like all the tests I have described, looks only at the initiation stage of carcinogenesis rather than other stages which are said to play an important role in tumour progression.

If methods could be devised to pick up pre-neoplastic changes early on, such as the production of embryonic antigens, then this type of test would obviously be of great advantage in the screening of compounds for potential carcinogenicity. Finally, there are methods that show that DNA damage has occurred *in vivo*; these methods employ density gradient centrifugation techniques of DNA from target organs but would appear at present to be too tedious for the routine screening of carcinogens (Cox *et al.* 1973). A procedure which may have some use in detecting the covalent reaction of carcinogen with DNA would be the monitoring of excreted carcinogen adducts in the urine of carcinogen treated animals, possibly by radioimmunoassay. This could be a sensitive technique provided the adducts were excreted unchanged and that the necessary antibodies could be raised.

Determination of priorities after carcinogen testing

Various schemes have been proposed to determine which assay should be carried out in any testing protocol so that priorities for long-term carcinogenicity tests can be determined (Bridges 1973; Flamm 1974; Bartsch 1976). Since carcinogens react with macromolecules, tests that depend on this for their function should be used. It is obvious from the foregoing sections that schemes should start off with the simple and move to the more complex. Thus mutation in microorganisms should be the initial assay system for routine screening of chemicals. Though the techniques are relatively simple, the results can often be complex. One knows the constituents of the assay and the end-point; however, the great number of factors which might be important in reaching the end-point of mutation should not be forgotten. These can include the role of bacterial metabolism in activation and detoxification of metabolites, aerobic oxidation of metabolites, the close proximity of bacteria and endoplasmic reticulum, the possibility of reactions of metabolites with the parent compound or other metabolites and the concentration of substrate and cofactors. Furthermore, it should be recognized that there are some chemicals which are active mutagens only if the assay is carried out in a particular manner. Thus, dimethylnitrosamine and diethylnitrosamine are only weakly or non-mutagenic in a standard plate test but active in a liquid suspension assay.

Accepting that bacterial mutagenicity is the simplest screening procedure we can use routinely, what other tests need to be done in conjunction with this assay to increase the reliability of predicting whether a particular chemical has carcinogenic potential? Studies in our laboratory indicate that as previously mentioned, the measurement of 'unscheduled DNA synthesis' in HeLa cells in culture with the addition of a liver enzyme preparation gives a reliable indication that a particular chemical is activated to an electrophilic species (Martin *et al.* 1978). It is my opinion that these two tests together give good predictive value for carcinogenic potential and can be used for pre-screening.

Compounds giving positive results in both tests should be viewed with suspicion. If the positive compound is part of a structural series, other members of which are negative and these have the desired activity, then naturally one should concentrate one's attention on the non-mutagenic compounds. If there is no alternative chemical or the benefits far outweigh the hazard, the necessary handling precautions should be taken to avoid heavy exposure of people employed in its manufacture. There are certain compounds for which, on the basis of their chemical structure, one can predict with some accuracy whether the compound might have carcinogenic potential. It has been suggested that in a short-term assay a structurally related compound of known carcinogenicity should be tested in conjunction with the test compound as a positive control. Naturally if the assay is negative for the known carcinogen it is unlikely to pick up activity from the test compound (Purchase *et al.* 1976).

Some compounds which are hormonal in their action, or which act through some physical characteristic such as fibre size, would have to be tested in animal studies immediately, without the necessity of doing a short-term test since most of these would not pick the compound up anyway. Chemicals such as diethylstilboestrol might be active in a short-term test but their action may not be related to their mechanism of initiating cancer. It is therefore pointless to try to find one or more assays that will detect every type of carcinogenic substance no matter what its mechanism of action.

CONCLUSIONS

The short-term tests that have been described here enable one to make a rational approach to the testing of chemicals for tumour initiating activity. They cannot, at the present time give any quantitative idea of hazard to man nor of carcinogenic potency (long-term animal studies may not be able to do this either). The tests can also provide a method of monitoring people for exposure to carcinogens by assaying urine extracts for mutagenicity. The use of the tests is in a toxicological testing programme for all compounds to which people are exposed. If closed processes are involved in manufacture, it is pointless testing these compounds unless individuals come into contact with them.

Finally it should be pointed out that there are many arguments over the predictive value of animal carcinogenicity tests. Most short-term test programmes have, however, depended heavily on the results of these to establish their predictive value. One hopes that this is not a question of two imponderables being added one to another.

REFERENCES (Garner)

Bartsch, H. 1976 Predictive value of mutagenicity tests in chemical carcinogenesis. *Mutat. Res.* **38**, 177–190.

Berwald, Y. & Sachs, L. 1963 In vitro cell transformation with chemical carcinogens. *Nature, Lond.* **200**, 1182–1184.

Bridges, B. A., Mottershead, R. P., Rothwell, M. A. & Green, M. H. L. 1972 Repair deficient bacterial strains suitable for mutagenicity screening: tests with the fungicide captan. *Chem.-biol. Interact.* **5**, 77–84.

Bridges, B. A. 1973 Some general principles of mutagenicity screening and a possible framework for testing procedures. *Environ. Hlth Persp.* **6**, 221–227.

Bridges, B. A. 1976 Short term screening tests for carcinogenicity. *Nature, Lond.* **261**, 195–200.

Brookes, P. & Lawley, P. D. 1964 Evidence for the binding of polynuclear aromatic hydro-carbons to the nucleic acids of mouse skin: relationship between carcinogenic power of hydrocarbons and their binding to deoxyribonucleic acids. *Nature, Lond.* **202**, 781–784.

Clayson, D. B. & Garner, R. C. 1976 Carcinogenic aromatic amines and related compounds. In *Chemical carcinogens* (American Chemical Society Monograph no. 173) (ed. C. E. Searle), pp. 366–461). Washington, D.C.: American Chemical Society.

Cleaver, J. E. 1970 DNA repair and radiation sensitivity in human (*Xeroderma pigmentosum*) cells. *Int. J. Radiat. Biol.* **18**, 557–565.

Cox, R., Damjanov, I., Abanoli, S. E. & Sarma, D. S. R. 1973 A new method for measuring DNA damage and repair in the liver *in vivo. Cancer Res.* **33**, 2114–2121.

DeMars, R. 1974 Resistance of cultured human fibroblasts and other cells to purine and pyrimidine analogues in relation to mutagenesis detection. *Mutat. Res.* **24**, 335–364.

DiPaolo, J. A., Nelson, R. L. & Donovan, P. J. 1971 Morphological oncogenic and karyological characterisation of Syrian hamster embryo cells transformed *in vitro* by carcinogenic polycyclic hydrocarbons. *Cancer Res.* **31**, 1118–1127.

DiPaolo, J. A., Nelson, R. L. & Donovan, P. J. 1972 *In vitro* transformation of Syrian hamster embryo cells by diverse chemical carcinogens. *Nature, Lond.* **235**, 278–280.

Doll, R. 1967 *Prevention of cancer: pointers from epidemiology.* Nuffield Provincial Hospitals Trust, The Rock Carling Fellowship.

Doll, R. & Vodopija, I. 1973 *Host environment interactions in the etiology of cancer in man.* (*I.A.R.C. Scientific Publication* no. 7.) Lyon, France: International Agency for Research on Cancer.

Doll, R. 1977 Strategy for detection of cancer hazards to man. *Nature, Lond.* **265**, 589–596.

Fears, T. R., Tarone, R. E. & Chu, K. C. 1977 False-positive and false-negative rates for carcinogenecity screens. *Cancer Res.* **37**, 1941–1945.

Flamm, W. G. 1974 A tier system approach to mutagen testing. *Mutat. Res.* **26**, 329–333.

Freeman, A. E., Weisburger, E. K., Weisburger, J. H., Wolford, R. G., Maryak, J. M. & Huebner, R. J. 1973 Transformation of chemicals as an indication of the carcinogenic potential of chemicals. *J. nat. Cancer Inst.* **51**, 799–807.

Garner, R. C., Miller, E. C., Miller, J. A., Garner, J. V. & Hanson, R. S. 1971 Formation of a factor lethal, for *S. typhimurium* TA1530 and TA1531 on incubation of aflatoxin B$_1$ with rat liver microsomes. *Biochem. biophys. Res. Commun.* **45**, 774–780.

Garner, R. C., Miller, E. C. & Miller, J. A. 1972 Liver microsomal metabolism of aflatoxin B$_1$ to a reactive derivative toxic to *Salmonella typhimurium* TA1530. *Cancer Res.* **32**, 2058–2066.

Garner, R. C. & Wright, C. M. 1973 Induction of mutations in DNA-repair deficient bacteria by a liver microsomal metabolite of aflatoxin B$_1$. *Br. J. Cancer* **28**, 544–551.

Garner, R. C. & Wright, C. M. 1975 Binding of [^{14}C]aflatoxin B$_1$ to cellular macromolecules in the rat and hamster. *Chem.-biol. Interact.* **11**, 123–131.

Garner, R. C. & Nutman, C. A. 1977 Testing of some azo dyes and their reduction products for mutagenicity using *Salmonella typhimurium* TA1538. *Mutat. Res.* **44**, 9–19.

Green, M. H. L. & Muriel, W. J. 1976 Mutagen testing using trp$^+$ reversion in *Escherichia coli. Mutat. Res.* **38**, 3–32.

Healey, P., Mawdesley-Thomas, L. E. & Barry, D. H. 1970 Short-term test for evaluating potential carcinogenic activity of tobacco condensates. *Nature, Lond.* **228**, 1006.

Heidelberger, C. 1973 Chemical oncogenesis in culture. *Adv. Cancer Res.* **18**, 317–366.

Huberman, E. & Sachs, L. 1974 Cell-mediated mutagenesis with chemical carcinogens. *Int. J. Cancer* **13**, 326–333.

I.A.R.C. Monographs on the evaluation of carcinogenic risk of chemicals to man, vols 1–16. Lyon, France: International Agency for Research on Cancer.

Kada, T., Tutikawa, K. & Sadaie, Y. 1972 *In vitro* and host-mediated 'rec-assay' procedures for screening chemical mutagens; and phloxine, a mutagenic red dye detected. *Mutation Res.* **16**, 165–174.

Magee, P. N. 1974 Testing for carcinogens and mutagens. *Nature, Lond.* **249**, 795–796.

Magee, P. N., Montesano, R. & Presusmann, R. 1976 *N*-nitroso compounds and related carcinogens. In *Chemical Carcinogens* (American Chemical Society Monograph no. 173) (ed. C. E. Searle), pp. 491–625. Washington, D.C.: American Chemical Society.

Malling, H. V. 1971 Dimethylnitrosamine: formation of mutagenic compounds by interaction with mouse liver microsomes. *Mutat. Res.* **13**, 425–429.

Martin, C. N. & Garner, R. C. 1977 Aflatoxin B-oxide generated by chemical or enzymic oxidation of aflatoxin B_1 causes guanine substitution in nucleic acids. *Nature, Lond.* **267**, 863–865.

Martin, C. N., McDermid, A. C. & Garner, R. C. 1977 Measurement of 'unscheduled' DNA synthesis in HeLa cells by liquid scintillation counting after carcinogen treatment. *Cancer Lett.* **2**, 355–360.

Martin, C. N., McDermid, A. C. & Garner, R. C. 1978 Testing of known carcinogens and noncarcinogens for their ability to induce unscheduled DNA synthesis in HeLa cells. *Cancer Res.* **38**, 2621–2627.

McCann, J., Choi, C., Yamasaki, E. & Ames, B. N. 1975 Detection of carcinogens as mutagens in the *Salmonella*/microsome test: assay of 300 chemicals. *Proc. natn. Acad. Sci. U.S.A.* **72**, 5135–5139.

Miller, J. A. 1970 Carcinogenesis by chemicals – an overview. (G. H. A. Clowes Memorial Lecture.) *Cancer Res.* **30**, 559–576.

Mohn, G. & Ellenberger, J. 1973 Mammalian blood mediated mutagenicity tests using a multipurpose strain of *Escherichia coli*, K-12. *Mutat. Res.* **19**, 259–260.

Mondal, S. 1975 Transformed liver cells obtained in culture from hepatectomised rats treated with dimethylnitrosamine. *Br. J. Cancer* **31**, 245–249.

Nikaido, O. & Fox, M. 1976 The relative effectiveness of 6-thioguanine and 8-azaguanine in selecting resistant mutants from two V79 Chinese hamster cells *in vitro*. *Mutat. Res.* **35**, 279–288.

Ponomarkov, C., Tomatis, L. & Turusov, V. 1976 The effect of long term administration of phenobarbitone in CF-1 mice. *Cancer Lett.* **1**, 165–172.

Ponten, J. 1976 The relationship between *in vitro* transformation and tumour formation *in vivo*. *Biochim. biophys. Acta* **458**, 397–422.

Purchase, I. F. H., Longstaff, E., Ashby, J., Styles, J. A., Anderson, D., Lefevre, P. A. & Westwood, F. R. 1976 Evaluation of six short term tests for detecting organic chemical carcinogens and recommendations for their use. *Nature, Lond.* **264**, 624–627.

Reuber, M. D. 1976 Various degrees of susceptibility of different stocks of rats to *N*-2-fluorenyldiacetamide hepatic carcinogenesis. *J. natn. Cancer Inst.* **57**, 111–114.

Rossi, L., Ravera, M., Repetti, G. & Santi, L. 1977 Long term administration of DDT or phenobarbital-Na in Wistar rats. *Int. J. Cancer* **19**, 179–185.

Saffiotti, U. & Wagoner, J. K. 1976 Occupational carcinogenesis. *Ann. N.Y. Acad. Sci.* **271**, 1–516.

San, R. H. C. & Stich, H. F. 1975 DNA repair synthesis of cultured human cells as a rapid bioassay for chemical carcinogens. *Int. J. Cancer* **16**, 284–291.

Shimkin, M. B., Weisburger, J. H., Weisburger, E. K., Gubareff, N. & Suntzeff, V. 1966 Bioassay of 29 alkylating chemicals by the pulmonary-tumour response in Strain A mice. *J. natn. Cancer Inst.* **36**, 915–935.

Shin, S. J. 1974 Nature of mutation conferring resistance to 8-azaguanine in mouse cell lines. *J. Cell Sci.* **14**, 235–251.

Sims, H. F. & Grover, P. L. 1974 Epoxides in polycyclic aromatic hydrocarbon metabolism and carcinogenesis. *Adv. Cancer Res.* **20**, 166–274.

Slater, E. E., Anderson, M. D. & Rosenkranz, H. S. 1971 Rapid detection of mutagens and carcinogens. *Cancer Res.* **31**, 970–973.

Stich, H. F. & San, R. H. C. 1973 DNA repair synthesis and survival of repair deficient human cells exposed to the K-region epoxide of benz(a)anthracene. *Proc. Soc. Exp. Biol. Med.* **142**, 155–158.

Stoltz, D. R., Poirier, L. A., Irving, C. L., Stich, H. F., Weisburger, J. H. & Grice, H. C. 1974 Evaluation of short-term tests for carcinogenicity. *Toxic. appl. Pharmac.* **29**, 157–180.

Suzuki, N. & Okada, S. 1976 Isolation of nutrient deficient mutants and quantitative mutation assay by reversion of alanine-requiring L5178Y cells. *Mutat. Res.* **34**, 489–506.

Thielmann, H. W. 1976 Carcinogen induced DNA repair in nucleotide permeable *Escherichia coli* cells. *Eur. J. Biochem.* **61**, 501–513.

Wattenberg, R. W., Loub, W. D., Lam, L. K. & Speier, J. L. 1976 Dietary constituents altering the responses to chemical carcinogens. *Fedn Proc. Fedn Socs exp. Biol.* **35**, 1327–1331.

Discussion

G. A. H. ELTON (*Ministry of Agriculture, Fisheries and Food, London SW1P 2AE, U.K.*). Environmental contaminants can affect man by various routes, one of which is via food. For example, for the average person, food is the main source of intake of lead and of mercury, and virtually the only source of intake of methyl mercury. Food additives and their reaction products with food components are also ingested.

In attempting to assess the risk, if any, to the population from additives and contaminants in food, we sometimes have assistance from guide-lines from W.H.O. and other medical authorities (e.g., for lead and for methyl mercury); these guide-lines are usually based on information about the toxicity for man of the material under consideration, taking into account industrial exposure, accidents, etc. For many substances, however, such information does not exist or is at best fragmentary, and we have to rely on information based on animal experiments and/or in-vitro studies. The interpretation of such information poses many problems, particularly in relation to possible carcinogenicity. Cancer risks can only be assessed in terms of probabilities; for example, at best one can only hope to be able to say that the 'average man (or woman)' exposed to compound A at $x\%$ in the diet for y years has a probability z of incurring cancer as a result. How can one make such an assessment for a population of many millions of people, eating varied diets, particularly when both x and z are very small? Studies on a few hundred animals are of limited use. To obtain positive results with a reasonable number of rats, for example, many research workers feed the test compound at levels up to the highest dose that the rat can tolerate (sometimes 5 % by mass of the diet or more). Extrapolation of the results for high dose levels to more realistic levels of ingestion (perhaps a few parts/10^6 or parts/10^9) is hazardous, especially when one is attempting to extrapolate almost to the origin of the dose–response graph. It is then still necessary to take the additional step of extrapolation from rat or other experimental animal to man.

These are fundamental problems of risk assessment. How far we can contribute to their solution by the use of short-term tests based on microorganisms or on mammalian cells remains to be seen. Meanwhile, risk assessment for additives and contaminants in food will continue to be based on skilled judgement by experts of data which are not always complete, but on which decisions must be taken to protect the public safety. Furthermore, we must always be prepared to revise our judgements at short notice if necessary, as valid new experimental evidence becomes available.

Proc. R. Soc. Lond. B. **205**, 135–143 (1979)

Printed in Great Britain

Cardiovascular disease and trace metals

By A. G. Shaper

Department of Clinical Epidemiology and Social Medicine, Royal Free Hospital, 21 Pond Street, Hampstead, London NW3 2PN, U.K.

Cardiovascular disease is a major cause of morbidity and mortality in the U.K. and other developed countries. In the U.K., mortality from coronary heart disease has increased progressively over the past 25 years, particularly in males. This paper examines the possible role of trace metals in the development of cardiovascular disease, with particular reference to the effects of cobalt, cadmium and lead in myocardial disease, atherosclerosis and hypertension. It is concluded that cobalt is an unimportant factor in community levels of cardiovascular disease, that cadmium has striking effects on blood pressure in animals and that there is some evidence for an association between environmental lead and high blood pressure.

Introduction

The concern of this symposium is with long-term hazards to man of man-made chemicals in the environment, and my task is to examine the effects of environmental chemicals on cardiovascular disease. In particular, I shall address myself to an apparently straightforward question: is there any convincing evidence that trace elements introduced into the environment by man play a significant role in the present burden of cardiovascular disease (c.v.d.) in the community? A wide range of trace elements has been linked with one or other aspect of c.v.d., some being regarded as beneficial to the heart and/or blood vessels and others as directly or indirectly harmful. This W.H.O. listing (see table 1) ranks those elements with reasonable evidence of possible involvement with c.v.d. as being of primary interest and the rest as being of secondary interest (Masironi 1974). My comments will be limited to those trace elements whose presence in excess in the

Table 1. Elements for which associations with cardiovascular disease are suggested (Masironi 1974)

(Bulk elements in italics.)

primary interest	secondary interest	
cadmium	lithium	strontium
chromium	fluorine	iodine
copper	*sodium*	mercury
selenium	silicon	lead
zinc	vanadium	
calcium	manganese	
magnesium	molybdenum	

environment is due to man's activities rather than to nature and where it is suspected that they play an important role in c.v.d. of numerical significance to society.

THE CARDIOVASCULAR SCENE

About one-half of all deaths in England and Wales are attributed to c.v.d. Coronary heart disease (c.h.d.) predominates and is almost invariably associated with atherosclerosis, a chronic and progressive thickening of the arteries supplying the heart muscle. Cerebrovascular disease (stroke) is the second major contributor to c.v.d. mortality, and both atherosclerosis and high blood pressure (hypertension) are important precursors of stroke.

In addition to the effects of atherosclerosis and high blood pressure, there are factors which appear to have a direct toxic action on the heart muscle. We may therefore have to consider whether trace elements affect myocardial function directly, as well as whether they affect atherosclerosis or the level of blood pressure.

In most populations, blood pressure rises with increasing age and the increasing levels of blood pressure are associated with increasing morbidity and mortality from cardiovascular disease. We do not know why blood pressure rises progressively with increasing age; we do know that this rise is not biologically normal.

Within the United Kingdom there is considerable regional variation in the frequency of cardiovascular disease. Death rates from c.v.d. are lowest in the southeast of England, where water is predominantly hard, and higher in the north and to the west, where drinking water is predominantly soft. Cardiovascular mortality rates are about 40 % higher in towns with very soft water compared with towns with very hard water (Gardner *et al.* 1969). There is evidence that blood pressure levels and heart rates (the latter possibly reflecting myocardial function) are higher in towns with soft water (Stitt *et al.* 1973), and that sudden death (again possibly reflecting a myocardial effect) may contribute to the excess mortality from c.h.d. in these towns (Crawford *et al.* 1977).

We do not know whether this observed relation between c.v.d. and water quality is a direct and causal one, or whether water quality is an indicator for some other set of factors. It has been suggested that the association between soft water and higher death rates from c.v.d. could be due to the aggressive action of soft and acid waters on the conducting systems of the water supply, leading to an increased concentration of some trace elements, in particular lead and cadmium. As the list of trace elements which have been incriminated in the development of c.v.d. is extensive, I shall only review those few for which there appears to be some direct evidence of an association between the trace element and c.v.d.

COBALT

The story of cobalt as a toxic element directly affecting the heart muscle is often used to illustrate the possibility that other trace elements, even more widely present in the environment, may contribute towards the present epidemic of cardiovascular disease. Cobalt appears to present a clear-cut example of a trace metal which, if taken in small amounts over a long period of time, may cause a form of heart muscle disease (Underwood 1975).

The cobalt story begins with the introduction of synthetic detergents and their extensive use in washing glassware. No matter how thoroughly rinsing is carried out, some detergent remains and this effectively prevents foam developing when tap beer is poured into the glass. To the beer drinker, this is just not acceptable and to restore and stabilize the foam, cobalt salts may be added to the beer after manufacture. In the early 1960s, this was done in the United States, Canada and Belgium (and possibly elsewhere), to be followed by several outbreaks of severe and often fatal heart failure in copious beer drinkers (Kesteloot *et al.* 1966; Morin *et al.* 1967; Alexander 1972). After some fascinating detective work, cobalt was eventually suspected and it was discovered that cobalt concentrations in the heart muscle of these subjects was ten times higher than in non-drinkers. When the use of cobalt salts was discontinued, no further outbreaks were reported. However, it soon became evident that the problem was not a straightforward one.

At the cobalt concentrations of 1–5 parts/10^6 used in the brewing process, it would require 24 pints of beer per day to supply about 8 mg of cobalt, an amount well below that which has been given without ill-effects for refractory anaemias of renal failure. Indeed, up to 300 mg/day has been used therapeutically without cardiotoxic effect. It seemed that cobalt toxicity by itself could not explain the severe myocardial failure and it was postulated that cobalt and high alcohol intakes were both necessary to induce the distinctive cardiomyopathy, plus an additional factor such as low dietary protein. It was also suggested that the calcium in the water used in brewing the beer might be a factor, so that even the water was not left out!

In animal experiments, oral cobalt administration of 20 mg kg^{-1} per day produced myocardial lesions strikingly similar to those observed in the beer-drinker's cardiomyopathy (Mohiudden *et al.* 1970). Cobalt plus ethanol failed to increase the incidence or severity of the disease but if animals were maintained on beer for 36 days before cobalt was given, the cardiotoxic effect of cobalt was enhanced (Wiberg *et al.* 1969). It seems likely that the patients with cobalt–beer cardiomyopathy had preexisting heart damage from chronic ethanol intake and malnutrition for months or years before cobalt was introduced and changed a slowly progressive disease into a fulminating one.

The moral of this story is that even though a trace metal is known to be deposited in the heart muscle where it is known to have a detrimental action, it may not be possible to demonstrate its mode of action in man and thus not possible to produce the evidence required for confidence in the role of the trace metal as a critical causal factor. There is no evidence that cobalt plays any role of importance in c.v.d. in this country and in no country is there evidence that it is more than a sporadic and unusual contributor to c.v.d. problems.

CADMIUM

Cadmium is probably the most extensively studied metal in relation to cardio-vascular disease and there is evidence of an association between cadmium and raised blood pressure (Anon. 1976). Much of the current controversy regarding this association relates to the question of which factor appears first. Does increased cadmium intake lead to high blood pressure and renal disease or vice versa? The classic descriptions of cadmium toxicity do not mention hypertension but refer to renal failure and bone disease, and even in Itai-Itai disease in Japan, apparently caused by cadmium pollution of water and rice, no excess of hypertension has been reported (Perry 1972).

Animal studies

Experiments on rats and dogs have repeatedly shown that cadmium given in extremely small amounts produces a rise in blood pressure (Schroeder 1964; Perry 1972; Thind et al. 1973) and that the use of chelating agents or zinc (which displaces cadmium from the renal binding protein) will lower the blood pressure (Schroeder & Buckman 1967). In the experimental animals it is apparently not the renal concentration of cadmium by itself that determines the rise in blood pressure, but rather the ratio of cadmium to zinc.

In man, cadmium levels show a higher concentration in the kidneys of hypertensives than in normotensives (Schroeder 1965), but it is difficult to be sure how far this finding is related to previous treatment for hypertension.

The Renfrew study

A recent study from Renfrew, Scotland, has examined the relation between blood cadmium and blood pressure in 70 hypertensive patients and 70 controls matched for age (45–64 years) and sex (see Beevers et al. 1976a). Most of the hypertensives (53/70) were on treatment and several (8/70) had evidence of renal failure.

There were no significant differences found in mean blood cadmium levels between hypertensive and normotensive groups in either sex. No differences were found between untreated hypertensives and patients treated with any single group of drugs, although it is of interest that the untreated group had a lower blood cadmium level than any one of the several treated groups. When all treated patients are combined into a single group, there is a highly significant difference in blood cadmium levels between treated and untreated groups. It appears that treatment may raise the blood cadmium level.

In both hypertensive and normotensive subjects, cigarette smokers had significantly higher blood cadmium levels than non-smokers and it is well established that cigarette smoking is a major source of cadmium intake (Lewis et al. 1972; Nandi et al. 1969). From several other studies, it is known that cigarette smokers do not have higher blood pressure levels than non-smokers.

TABLE 2. BLOOD CADMIUM LEVELS IN TREATED AND UNTREATED
HYPERTENSIVE SUBJECTS (BEEVERS *ET AL.* 1976*a*)

		blood cadmium (μg/l)
untreated hypertensives	(17)	1.74
treated hypertensives	(71)	2.35
diuretics	(28)	2.34
beta-blockers	(20)	2.34
adrenergic blockers	(22)	2.32

The Kansas study

Kansas City, Kansas, and Kansas City, Missouri, are on opposite sides of the Missouri River and draw their water from the same source, but the water for Missouri is softened to about half the total hardness found in Kansas. The mortality rate from c.v.d. in Kansas (hard water) is significantly higher than in Missouri (soft water), the reverse of that which occurs in many large national studies. Volunteers (260 pairs of subjects) from the two cities, matched for age, sex, socio-economic status, smoking and body build were studied, and single water samples were collected from 10 % of the volunteers (25 in each city) (Bierenbaum *et al.* 1976). The tap water studies showed that Kansas water contained three times as much cadmium as Missouri tap water (3.0 compared with 1.0 μg/l) and one-quarter the concentration of zinc (4.0 compared with 16·0 μg/l).

The Kansas subjects (higher c.v.d. rates, hard water) had significantly higher blood pressures and more electrocardiographic abnormalities. They had higher blood concentrations of cadmium, lithium, magnesium, sodium, potassium and calcium as well as higher blood sugar and uric acid levels. Chromium, copper and zinc concentrations were lower than in Missouri. From this complex mass of data, the authors selectively conclude that the higher c.v.d. mortality is associated with increased hypertension, and with increased serum cadmium and decreased serum zinc concentrations. They focus on the cadmium:zinc ratio as the possible toxic factor in the drinking water.

The paper has been strongly (and I think justifiably) criticized by workers from the National Institutes of Health, U.S.A., who question the validity of the measurements and the extreme differences in cadmium and zinc serum levels, and who regard the statistical treatment of the data as inappropriate (Sharrett & Feinleib 1976). They conclude: 'We feel that the major conclusions of the study reflect more a point of view than a scientific result.' In all fairness, it seems possible that Kansas City, Kansas, may have a cadmium problem but that this is responsible for the difference in blood pressure distribution between the two sample groups cannot be regarded as established.

In the studies relating cadmium to c.v.d., it is the animal experimental work that is most striking. Cadmium administered in small doses to rats will raise their blood pressure, especially if the cadmium administration is combined with a soft water régime. In humans, the relation between raised blood pressure and c.v.d. is not convincing, and carefully designed studies are required before we can raise cadmium to the ranks of a serious contender for responsibility in the aetiology of raised blood pressure. Cigarette smoking is clearly associated with increased cadmium intakes and raised blood cadmium levels, and it is a major risk factor for c.v.d. Thus cadmium may play an important role in c.v.d. or may merely be a marker of cigarette smoking.

LEAD

There is reasonable evidence that clinical (symptomatic) lead poisoning may include signs suggestive of a toxic action on the heart, both in adults and in children, but little convincing evidence that smaller degrees of exposure to lead are associated with cardiovascular disease. Experimentally, it has not been possible to associate hypertension or heart disease with lead ingestion and perhaps the major reason for lead remaining linked with c.v.d. in current thinking is the presence of elevated lead levels in many soft drinking waters and the association of soft water with higher c.v.d. death rates. The acid nature of soft water leads to increased plumbosolvency, so that in some areas, e.g. Scotland, an important proportion of tap water samples may contain high concentrations of lead.

Recent studies in the Renfrew area of Glasgow have compared blood and tap water lead levels in hypertensive and normotensive subjects and concluded that in the West of Scotland, high blood pressure is associated with increased blood lead concentrations and that this might explain the high prevalence of cardiovascular disease in the area (Beevers et al. 1976b).

In the Renfrew study, 135 hypertensive subjects (74 males, 61 females) aged 45–64 years were compared with a similar number of normotensive subjects. There were no significant differences in mean blood lead levels between hypertensive and normotensive subjects in either males or females, although levels were slightly higher in hypertensive subjects in both sexes. A comparison of blood lead levels in paired hypertensive and normotensive male subjects (with arbitrary cut-off points for the blood lead levels) yielded a significant excess of pairs in which the hypertensive subjects had higher blood lead levels than the normotensive subjects. This excess was not seen in the female subjects. This finding in the male subjects is the basis for the conclusion regarding the association between blood lead and hypertension. Either one must regard hypertension in females as a different problem, or one must regard the lack of replication in the female group as diminishing the strength of the relation between lead and hypertension. Although the 5% level of significance is by convention regarded as meaningful, in an investigation with implications as critical as this, one would have welcomed replication with additional normotensive control groups from within the same large study in the Renfrew population.

The same group of workers from Glasgow have examined the possible harmful effects, especially on renal function, of long-term exposure to lead in the domestic water supply of people in predominantly rural districts, people selected because their homes were thought to have lead pipes or storage tanks (Campbell *et al.* 1977). Of the 970 selected households studied, 219 (23 %) had lead levels greater than 100 µg/l (0.48 µmol/l), the present W.H.O. limit for drinking water. Blood pressures and blood samples were obtained for 283 subjects from 136 of the 219 households. With the use of 54 age-matched and sex-matched pairs of subjects (male and female) with normal serum urea levels and with 'uraemia' (defined as a serum urea above 6.6 mmol/l; 50 mg/100 ml), a significant association was demonstrated between raised blood lead and raised serum urea concentrations. Similarly, by using 20 pairs of age-matched and sex-matched subjects (male and female) with raised and normal serum urate concentrations, a significant association was demonstrated between raised blood lead and raised serum urate concentrations.

Although the age distribution of the subjects in this study is not provided, there is evidence in the paper that this is a somewhat elderly group. Thus, hypertension was found in 31 subjects, of whom only 6 were aged 45–64 years and the rest were presumably above this age. Elderly subjects often exceed the serum urea level herein defined as 'uraemia' and no indication is given of the frequency with which this level was exceeded. Thus many of the 'uraemic' subjects in the matched pairs may have had serum urea levels within the range accepted as 'normal' for elderly persons, bringing into question the concept of the 'cut-off' point and the subsequent matching procedure (Hodkinson 1977). Furthermore, as in the Renfrew study (Beevers *et al.* 1976*b*), arbitrary cut-off points were used for blood lead levels, which may not be the most unbiased method for examining the data.

Blood lead concentrations in the 31 hypertensives did not differ significantly from those of an age and sex-matched control group. The blood pressures were below 90 mmHg (*ca.* 12 kPa) in all of the 'uraemic' subjects and the serum urea levels were below 40 mg/100 ml in all of the hypertensive subjects. These observations are interesting but one pauses for thought in a study in which the 283 blood pressures were taken by 37 different general practitioners. In retrospect, it would have been more informative to have examined matched subsamples of those with water lead levels below as well as above the cut-off point used (100 µg/l; 0.48 µmol/l). Differences in the frequency of uraemia (as defined), hypertension and hyperuricaemia would then certainly have been more meaningful. Despite intuitive feelings about the harmfulness of lead, firm evidence that 'excessive lead in domestic water supplies has a harmful effect on the community's health' as far as cardiovascular and renal disease is concerned, is still awaited.

Conclusion

Some of the evidence linking trace metals to cardiovascular disease has been examined and I must conclude on a note of uncertainty, in full awareness of the complexity of the factors underlying the development of each of the many cardiovascular disorders.

Cobalt can produce myocardial damage under certain circumstances, but it appears to be a sporadic and unimportant contributor to the community levels of c.v.d. Attempts to define its mechanisms of action in humans illustrate the considerable complexity which presents when attempting to focus on the specific and possibly independent role of any single trace metal. Cadmium has striking effects on blood pressure in animals. In humans, the epidemiological evidence linking cadmium to hypertension is limited and possibly non-existent, although the levels of cadmium in cigarette smokers provides an aura of guilt by association. Lead has emerged from the Scottish studies as weakly suspect in the aetiology of raised blood pressure and/or renal disease. The high prevalence of plumbosolvent soft waters in Great Britain and their association with high levels of c.v.d. mortality makes the study or lead and c.v.d. of considerable importance. We must remain aware that blood lead or water lead may merely be acting as an indicator of other sets of factors which are independently associated with the blood pressure story.

REFERENCES (Shaper)

Alexander, C. S. 1972 Cobalt–beer cardiomyopathy. *Am. J. Med.* **53**, 395–417.

Anon. 1976 Cadmium, lead and hypertension. *Lancet* ii, 1230–1231.

Beevers, D. G., Campbell, B. C., Goldberg, A., Moore, M. R. & Hawthorne, V. M. 1976*a* Blood-cadmium in hypertensives and normotensives. *Lancet* ii, 1222–1224.

Beevers, D. G., Erskine, E., Robertson, M., Beattie, A. D., Campbell, B. C., Goldberg, A., Moore, M. R. & Hawthorne, V. M. 1976*b* Blood-lead and hypertension. *Lancet* ii, 1–3.

Bierenbaum, M. L., Fleischman, A. I., Dunn, J. & Arnold, J. 1976 Possible toxic water factor in coronary heart-disease. *Lancet* i, 1008–1010.

Campbell, B. C., Beattie, A. D., Moore, M. R., Goldberg, A. & Reid, A. G. 1977 Renal insufficiency associated with excessive lead exposure. *Br. med. J.* i, 482–485.

Crawford, M. D., Clayton, D. G., Stanley, F. & Shaper, A. G. 1977 An epidemiological study of sudden death in hard and soft water areas. *J. chron. Dis.* **30**, 69–80.

Gardner, M. J., Crawford, M. D. & Morris, J. N. 1969 Patterns of mortality in middle and early old age in the county boroughs of England and Wales. *Br. J. prev. Soc. Med.* **23**, 133–140.

Hodkinson, H. M. 1977 *Biochemical diagnosis of the elderly.* London: Chapman & Hall.

Kesteloot, H., Terryn, R., Bosmans, P. & Joossens, J. V. 1966 Alcoholic perimyocardiopathy. *Acta cardiol.* **21**, 341–357.

Lewis, G. P., Jusko, W. J. & Coughlin, L. L. 1972 Cadmium accumulation in man: influence of smoking, occupation, alcohol habit and disease. *J. chron. Dis.* **25**, 717–726.

Masironi, R. (ed.) 1974 *Trace elements in relation to cardiovascular diseases.* Geneva: World Health Organization.

Mohiudden, S. M., Taskar, P. K., Rheault, M., Roy, P.-E., Chenard, J. & Morin, Y. 1970 Experimental cobalt cardiomyopathy. *Am. Heart J.* **80**, 532–543.

Morin, Y. L., Foly, A. R., Martineau, G. & Roussel, J. 1967 Quebec beer-drinking cardiomyopathy: forty-eight cases. *Can. med. Ass. J.* **97**, 881–883.

Nandi, M., Jick, H., Slone, D., Shapiro, S. & Lewis, G. P. 1969 Cadmium content of cigarettes. *Lancet* ii, 1329–1330.

Perry, H. M. 1972 Cardiovascular diseases related to geochemical environment. *Ann. N.Y. Acad. Sci.* **199**, 202–216.

Schroeder, H. A. 1964 Cadmium hypertension in rats. *Am. J. Physiol.* **207**, 62–66.

Schroeder, H. A. 1965 Cadmium as a factor in hypertension. *J. chron. Dis.* **18**, 647–656.

Schroeder, H. A. & Buckman, J. 1967 Cadmium hypertension. Its reversal in rats by a zinc chelate. *Archs envir. Hlth* **13**, 788–789.

Sharrett, A. R. & Feinleib, M. 1976 Possible toxic factor in coronary heart-disease. *Lancet* ii, 76.

Stitt, F. W., Crawford, M. D., Clayton, D. G. & Morris, J. N. 1973 Clinical and biochemical indicators of cardiovascular disease among men living in hard and soft water areas. *Lancet* i, 122–126.

Thind, G. S., Biery, D. N. & Bovee, K. C. 1973 Production of arterial hypertension by cadmium in the dog. *J. Lab. clin. Med.* **81**, 549–556.

Underwood, E. J. 1975 Cobalt. *Nutr. Rev.* **33**, 65–69.

Wiberg, G. S., Munro, I. C., Meranger, J. C., Morrison, A. B., Grice, H. C. & Heggveit, H. A. 1969 Factors affecting the cardiotoxic potential of cobalt. *Clin. Toxic.* **2**, 257–271.

Discussion

G. KAZANTZIS (*Department of Community Medicine, The Middlesex Hospital, London, W1, U.K.*). Industrial workers have had heavier exposure to cadmium in the past than any comparable general population group. It would be expected, therefore, that if cadmium absorption gave rise to hypertension in man, this would be evident in cadmium workers. While no extensive epidemiological study of hypertension in these workers has yet been performed, the available evidence does not suggest the existence of a hypertensive effect. Hypertension has not been prominent as a clinical finding in workers with chronic cadmium poisoning, and careful observations made in Sweden in cadmium workers have shown no excessive prevalence of hypertension. Professor Shaper mentioned renal disease as a feature of chronic cadmium poisoning, which may suggest the presence of an associated hypertension. However, the renal disorder seen in this condition is a tubular dysfunction and does not primarily affect the glomeruli. It is thus not the type of renal disorder associated with either primary or secondary hypertension.

R. SCHOENTAL (*Department of Pathology, Royal Veterinary College, London, U.K.*). Professor Shaper's account of the incidents which appeared to incriminate cobalt in beer in relation to cardiovascular disease might possibly be explained by the presence of T-2 toxin in beer. Mouldy brewers' grains, containing T-2 toxin have been responsible for the death from a haemorrhagic syndrome of 9 out of 115 dairy cows in Scotland during the winter 1976/77 (Petrie *et al.* 1977 *Vet. Rec.* **101**, 326 (1977); Schoental 1977 *Vet. Rec.* **101**, 473). T-2 toxin, if occasionally present in barley (e.g. after wet harvest) would be extracted by alcohol into beer; its concentration in beer could vary greatly in various batches, and explain the difficulty in correlating cardiovascular disease with drinking of alcoholic beverages.

Proc. R. Soc. Lond. B. **205**, 145–151 (1979)

Printed in Great Britain

Moderately raised blood lead levels in children

By R. Lansdown

*Department of Psychological Medicine, The Hospital for Sick Children,
Great Ormond Street, London WC1N 3JH, U.K.*

There is no doubt that high blood lead levels are associated with mental subnormality and hyperactivity. Several recent studies in Britain and America have investigated the relation between moderate levels, i.e. between 20 and 40 µg/100 ml and behavioural and cognitive phenomena. Epidemiological studies have generally failed to point to a clearcut relation between such levels and overactivity or decrements in performance on standard intelligence and educational tests. Published studies with the use of chelation techniques have suffered from methodological weaknesses. It is known that socio-economic factors are powerfully related to measured intelligence and behaviour and, on the evidence available, it is to them that attention should primarily be given if preventative measures are being considered.

There remains the possibility that more refined test measures would detect impaired functioning in children with moderately raised lead levels, and that there is an interaction effect between lead and host resistance.

The subject of this paper carries a high emotional charge. 'Lead poisoning' is a loaded phrase and it is easy to arouse emotions with descriptions like 'one of the oldest and stealthiest poisons known to man' (Ottaway & Terry 1976).

The effects of damaging levels of lead ingestion, notably behaviour disturbance and learning difficulty, are also high arousal subjects. Those of us who have been concerned with educational failure frequently find ourselves under pressure to provide a single, testable answer to the question 'Why can't Johnny read?' and the history of education is full of attempts to meet that pressure. Lead, for some commentators, appears in part to meet it.

No one writing in the last few years has doubted that high levels of lead in the body can be harmful. As Millar *et al.* (1970) noted, the health hazard has been discussed since the seventeenth century. But, as the report of the Department of the Environment Working Group on Heavy Metals made clear, lead is naturally present in all components of our environment and in man himself (Department of the Environment 1974). In a study published in 1967, quoted in the Working Group's report, the average blood lead level for U.K. inhabitants was 23 µg/100 ml, which was exactly the same figure as that obtained from a sample in Central Brazil and only 1 µg/100 ml higher than the figure gained from mountain dwellers in New Guinea. The D.O.E. report concludes that 'evidence of harmful effects in adults is rarely seen at blood levels below 80 µg/100 ml; indeed, cases of poisoning usually involve levels considerably in excess of this'.

Symptoms in children are different from those in adults and the disease is generally more serious, though signs may appear at lower blood lead concentrations. The report underlines this conclusion in its next paragraph, which cites levels up to 40 µg/100 ml as normal; from 40 to 80 µg as acceptable, from 80 to 120 µg/100 ml as excessive and more than 120 µg/100 ml as dangerous.

These figures, though, are described as both arbitrary and not applicable to children. As with adults, there is no doubt that a high level of lead, say 120 µg/ 100 ml, is dangerous and can give rise to behavioural and learning difficulties. There is, however, a grey area, of around 25–60 µg/100 ml, where there is considerable uncertainty about the effects of lead on children. This moderately raised level is the subject of my paper today. I propose first to report some studies which have argued that a moderate level results in disturbance of functioning, and then others, including an epidemiological study, which found no such thing. This will be followed by an interpretation of later work which, in my view, supports the null hypothesis, and I shall conclude with a mention of a recent work which may stand my argument on its head.

The work of David and his colleagues in America has consistently pointed to the adverse effects of very moderately raised lead levels (David et al. 1972, 1976). For example in the 1972 study they reported on 91 children who had been referred to the Out Patient Clinic of a Brooklyn hospital. Of these, 54 were placed in a 'hyperactive' group and 37 in the control group. The former had blood lead levels of 26 µg/100 ml compared with 22 µg/100 ml in the controls. There were also differences in urine lead levels in the same direction. Now, what can be concluded from these figures? Both the hyperactive and the control group have blood lead levels very close to the already quoted average for adults. With no evidence to the contrary it would appear that David's two samples came from a population exposed to roughly the same amount of lead and one must, therefore, look for other possible causes for the increased hyperactivity of the 54 children. It is possible that disturbance preceded lead ingestion, leading to inappropriate feeding habits. The general possibility of such an antedating is acknowledged by Landrigan et al. (1975) and remains a plausible explanation.

Evidence against the simple equation of raised lead levels equals adverse effects comes from Kotok (1972) in which 24 children with raised lead levels were compared with two comparison groups. Both the experimental and comparison groups included children with delays in fine motor and language areas and there was no evidence for a relation between raised lead levels and the results of the developmental test used. There was, however, a correlation between the degree of environmental disruption experienced by the children and their developmental scores. Kotok also quotes unpublished work by H. Costoff and S. Provence which yielded similar results, finding a correlation between development and the quality of mother–child relationships and none between development and lead levels.

There was, then, no clearcut evidence on what I have already referred to as this grey area when in 1972 my colleagues and I became aware of the population living in the vicinity of a smelting works. The factors causing contamination had been present for about 30 years, so we could assume that all children born and still living in the area had been exposed to a higher than normal level of environmental lead all their lives. The intake of lead was derived in part by inhalation of airborne emissions, in part by inhalation of dust in houses and on roads and in part by the ingestion of dusts.

The area is not one of London's most fashionable quarters. A description of it in 1972 by a student of London history saw it as '. . . a desperate place . . . every indicator of urban health is pointing resolutely downwards . . . shops are woefully inadequate . . . factories are closing and jobs fleeing . . . the housing is mostly unattractive and unpopular – 97 % of it being in public ownership. At one stage there were 400 council flats untaken and the council admit that the "housing officer is obliged to go further down the waiting list than is normal" to fill them . . . it is hard to see why any young people should stay'. I have quoted at some length from this article because much of the interpretation of our figures rests on a knowledge of the area as a social entity as well as just a place where children gained high lead levels.

Here we had a densely populated area in which we felt we could examine the results of moderately raised lead levels epidemiologically, i.e. we did not take a selected sample with all the bias that such a technique can give rise to; we were able to have access to a total population. This methodological point is stressed since it is comparatively rare in research of this kind. The population to be studied was identified by a house-to-house search for children living within 500 m of the works. A total of 476 under the age of 16 years were found, of whom 119 were under the age of 5 years.

Lead levels were determined in duplicate in 10 µl samples of whole blood by atomic absorption spectrophotometry (Delves 1970). Recent work has shown that the micro-method we used may overestimate the lead in blood by 5 µg/100 ml. Since much of our statistical analysis relied on correlation techniques the uncertainty of our measures is less important than it might otherwise have been.

Our first question was whether there was evidence of raised lead levels and whether such levels seemed to be related to the works in question. The mean lead level of the children in the main study, i.e. those between 6 and 16 years ($n = 125$) was 33 µg/100 ml with a standard deviation of 9. Exactly 20 % had levels of 40 or over. The figure for pre-school children with a 40+ level was a similar 21 %. More striking was the relation between lead levels and proximity of the children's home to the works. Taking mothers, pre-school and school age children separately there was a statistically significant relation, with the trend of nearer to the works meaning higher lead levels. (See Lansdown *et al.* (1974) for details.)

The next question was to examine the effect of such exposure on children's functioning. An examination of the literature suggested that we should look at several areas, including:

(a) intelligence, if possible using tests which would differentiate between verbal skills and those involving visuo-motor abilities;

(b) general behavioural characteristics, with particular reference to over-activity;

(c) reading ability.

Because we were fortunate enough to be able to carry out a population study we did not need a control group but, since we were planning to employ correlational techniques we did need tests which fulfilled the requirements of parametric statistics, and we therefore chose an established, well standardized test to look at intelligence, namely the Wechsler Scales for Children and Adults. Because of the standardization characteristics of these scales we could not test all of the pre-school children and therefore chose the age group from 6 to 16 years.

The reading test used was the Burt Graded Word Test. This is simple to administer and has a highly significant correlation with other standardized tests (Lansdown 1973). All testing was carried out by supervised psychology graduates.

Information of the previous addresses of families, the length of stay in any one address, the social class of the father (based on occupation) and length of education of both parents was obtained by health visitors who made home visits to each family.

Ratings of behaviour were gained in school, by using a scale filled in by teachers (Rutter 1967). This scale is of known reliability and validity and has been widely used both in primary school (Rutter 1973) and in secondary schools (M. J. Rutter, personal communication). It contains three items relevant to overactivity and enabled us to extract an admittedly crude measure of this characteristic.

The results of our tests can be stated fairly simply: they were negative. To take intelligence first: The mean I.Q. was 101 with a standard deviation of 11. In order to look more closely at these figures we correlated them with blood lead levels and found a correlation of 0.05, i.e. there was no significant relation.

Behavioural ratings gave results which were, to some people, astonishingly high. Overall 22.8 % of the children were rated by their teachers as disturbed. This is a high figure only to anyone not acquainted with London schools in general and socially deprived areas in particular. I recently spoke at a conference of teachers in Somerset and heard a London teacher introduced as 'straight from the battle-field'. Rutter, in a study of children 10 years old in another London borough, used the same scale and gained a total of 23.2 % rated as disturbed (Rutter 1973).

The children seen as disturbed were considered as a function of their lead levels and again there was no significant relation. The same was found when overactivity was considered, and an examination of reading scores yielded similar results.

There was, however, an anomalous result. When we broke down our figures by lead levels, we found that children with higher levels had, generally, higher intelligence scores and a lower rate of disturbance. This we found difficult to comprehend until we plotted scores with geographical mobility in mind. It was pointed out to us that much of the new housing in the district was on the extreme edge of the area that we surveyed. This was the housing that the local council found so difficult to fill because the district is so unfashionable.

An examination of test results of children according to where they had spent the first two years of life shed some light on our apparently anomalous finding, since there emerged a new equation: the nearer the lead works meant the higher the lead level, which meant the more stable the family, which meant the higher intelligence. When we isolated children who had not spent the first two years of their lives in the area this pattern became clear. Anecdotal evidence from residents and health visitors bore out the comments of the Housing Officer quoted above, i.e. the newcomers were those who might be expected to produce less intelligent, more disturbed children for reasons quite other than lead ingestion. There was, however, one slight hint that our findings were not as simple as that. The pattern of overactivity in children showed the same general picture, with the 'newcomers' being more overactive, but there was a trend among the others which suggested that the nearer the works one was born the higher would be one's overactivity. This failed to reach statistical significance but it should be noted.

Several other reports have been published since this one. Landrigan *et al.* (1975) examined 46 children with lead levels from 40 to 68 μg/100 ml (mean 48) and compared them with 78 controls. All children lived within 6.6 km of a lead emitting smelter. They conclude that 'chronic lead absorption . . . may produce subclinical impairment in neuropsychologic function'. I question, though, the grounds for going beyond the tentative 'may' in this conclusion. There were no significant differences between the two groups on 18 out of the 21 measures taken, including hyperactivity and full scale intelligence quotient. Of the three tests yielding different scores one can be explained as a statistical artefact and the other two have internal contradictions, being similar in content to two which were not different.

In a large-scale study, Hebel *et al.* (1976) examined the 11-plus scores of 851 Birmingham children living since birth in a lead polluted area and compared them with 1642 children from two similar but unpolluted areas. They found that children living in lead polluted communities were academically more successful than those living in unpolluted communities. Their conclusion was that the quality of education in the schools in the lead area was higher than in the other two and that this factor was paramount in explaining their results.

There have been other related studies, e.g. on water borne lead (Beattie *et al.* 1975), but my own conclusion so far remains that epidemiological evidence suggests that the effects of moderately raised lead levels in children have been exaggerated. It is, of course, possible that tests used so far have been too coarse to pick up the fine differences that do exist. It is possible, as P. Bryce-Smith has suggested (personal communication), that there is an interaction between lead and some other factor in the child. Both points need evidence to substantiate them. Failing such evidence, I suggest that we should look primarily to social and educational measures if we wish significantly to improve children's learning and behaviour.

There remains one final piece of work, that of the de-leading chelation procedure used by David and his colleagues, reported last year (David *et al.* 1976). To summarize: David took a group of hyperactive children, found that some had levels of 25 µg/100 ml or more and de-leaded them by using chelation medication. In those children for whom there was no other known cause for hyperactivity there was a dramatic improvement in behaviour. Unfortunately the number of children involved was small, only 13 in the whole study and there was no report of the use of a placebo. David concludes by admitting that many children have raised lead levels and are asymptomatic and looks to host resistance as a major factor in what he describes as a complex interplay. I understand that a report of a placebo trial is in preparation (D. Bryce-Smith, personal communication).

I said at the beginning of this paper that this final work, on chelation, might stand my own conclusions on their head. This is not strictly true, of course, because the emphasis on interaction effects is quite different from the earlier quoted work which sought to show that a moderately raised lead level is, *ipso facto*, dangerous. Given that human behaviour and learning are multifactorial in their determinants, it seems likely that an approach allowing for this characteristic will possibly be fruitful in future.

REFERENCES (Lansdown)

Beattie, A. D., Moore, M. R., Goldberg, A., Finlayson, M. J. W., Graham, J. F., Mackie, E. M., Main, J. C., McLaren, D. A. & Stewart, G. T. 1975 Role of chronic low level lead exposure in the aetiology of mental retardation. *Lancet* ii, 580–592.

David, O., Clark, J. & Voeller, K. 1972 Lead and hyperactivity. *Lancet* ii, 900–903.

David, O., Hoffman, S. J., Sverd, J., Clark, J. & Voeller, K. 1976 Lead and hyperactivity. Behavioural response to chelation: a pilot study. *Am. J. Psychiat.* **131**, 1155–1157.

Delves, H. T. 1970 A micro-sampling method for the rapid determination of lead in blood by atomic-absorption spectrophotometry. *Analyst, Lond.* **95**, 431–438.

Department of the Environment 1974 *Lead in the environment and its significance to man.* (Pollution Paper no. 2.) London: H.M.S.O.

Hebel, J. R., Kinch, D. & Armstrong, E. 1976 Mental capacity of children exposed to lead pollution. *Br. J. prev. soc. Med.* **30**, 170–174.

Kotok, D. 1972 Development of children with elevated blood lead levels: a controlled study. *J. Pediat.* **80**, 57–61.

Landrigan, P. J., Whitworth, R. H., Baloh, R. W., Stachling, N. W., Barthel, W. F. & Rosenblum, B. F. 1975 Neuropsychologic dysfunction in children with chronic low-level lead absorption. *Lancet* i, 708–712.

Lansdown, R. G. 1973 A study of the effect of severe visual handicap on the development of some aspects of visual perception and their relationship to reading and spelling in special schools for the partially sighted. Ph.D. thesis, C.N.A.A., London.

Lansdown, R. G., Clayton, B. E., Graham, P. J., Shepherd, J., Delves, H. T. & Turner, W. C. 1974 Blood lead levels, behaviour and intelligence: a population study. *Lancet* i, 538–541.

Millar, J. A., Cumming, R. L. C., Battistini, V., Carswell, F. & Goldberg, A. 1970 Lead and d-aminolaevulinic acid dehydratase levels in mentally retarded children and in lead poisoned suckling rats. *Lancet* ii, 695–698.

Ottoway, M. & Terry, A. 1976 The great lead petrol lie. *Sunday Times*, 11 January.

Rutter, M. J. 1967 A children's behaviour questionnaire for completion by teachers. *J. Child Psychol. Psychiat.* **8**, 1–11.

Rutter, M. J. 1973 Why are London children so disturbed? *Proc. R. Soc. Med.* **66**, 1221–1225.

Proc. R. Soc. Lond. B. **205**, 153–156 (1979)

Printed in Great Britain

Polybrominated biphenyls in Michigan [abstract only]

By I. J. Selikoff

*Mount Sinai School of Medicine, City University of New York,
Fifth Avenue and 100th Street, New York, N.Y. 10029, U.S.A.*

Inadvertent admixture of the fire retardant chemical polybrominated biphenyl (PBB) with farm feed occurred in Michigan late in 1973; important animal disease resulted. Delay in detection allowed wide and prolonged distribution of contaminated food, and as a result most of Michigan's 9 000 000 citizens currently have detectable levels of PBB in serum and body tissues. Almost all samples of breast milk in the State have been found to have PBBs as well as PCBs. A clinical field survey of more than 1000 dairy farm residents and others demonstrated an unusual prevalence of specific but not pathognomonic clinical complaints, particularly neuro-behavioural. These were far fewer among a control group of Wisconsin dairy farm residents not exposed to PBB. Similar contrasts were noted in two laboratory studies: liver function test results (enzyme levels) were much more likely to be abnormal in Michigan than in Wisconsin, possibly related to the fact that the halogenated biphenyl is a powerful enzyme inducer; in addition, T cell and B cell abnormalities and impaired lymphoblastogenesis were the rule in Michigan dairy farmers, but these findings, consistent with immunosuppression, were absent in the Wisconsin group. The long-term significance of the observations is not at present known. This is being studied, as is the status of the general population of Michigan presumed to have lower tissue burdens than the farmers.

Discussion

Elizabeth McLean (*Department of Psychiatry, St George's Hospital Medical School, London, U.K.*). Fatigue, weight loss and memory loss are all cardinal symptoms of depression. Classically, depression follows an important loss, not only the loss of a loved human figure, but also loss of self-esteem or loss of success at work. In particular, one would expect a farmer who loses his stock to be at high risk of a depressive illness.

It appears that no one has looked at the Michigan data from this point of view. This perspective casts a light on the absence of any relation between plasma levels and symptoms, and absence of symptoms in the workers making PBBs in the factory.

R. Peto (*Radcliffe Infirmary, Oxford, U.K.*). There is an obvious possibility in a study such as that reported by Professor Selikoff that, because of the public alarm about PBBs in Michigan since 1973–74, people who know that they have been exposed to PBBs will report symptoms such as sleep disturbance, depression, weakness, fatigue and poor memory even if the doses of PBBs were without real effect. To exclude such an artefact as the explanation of his data, Professor Selikoff must not merely compare people who live on, or who ate produce from, quarantined farms with other people who were not associated with quarantined farms, he must demonstrate that among people living on quarantined farms, a dose–response relation exists between serum (or adipose) PBB levels and the probability of the various neurological signs and symptoms which he has enquired about. If no clear relation exists, then it is probable that severe biases affect his data.

I. J. Selikoff. Both Dr McLean and Mr Peto are correct in pointing to difficulties associated with the interpretation of subjective symptoms. The problem of differentiating 'psychosomatic' symptoms from those associated with organ dysfunction may be difficult. Yet, despite uncertainty, symptoms have often been the initial clues directing clinical differential diagnosis.

Epidemiologically, we feel most secure when a clear statistical dose–response relation (exposure symptom) can be demonstrated. It must always be kept in mind that the yardstick measuring the 'cause' may be inappropriate.

In the presence of inadequate understanding, we fall back upon the hoped-for security of a dose–response relation. The 'dose' is presumed to be related to serum levels of the chemical, which is what we can measure. We cannot measure concentrations in the livers of exposed people, or in their brain cells, or in the membranes of their lymphocytes, though these might be more pertinent. We can biopsy subcutaneous depot fat, but not the thymus or adrenal, in our studies of this lipid-soluble halogenated hydrocarbon, and hope that what we examine may be relevant.

Animal tests tell us that hepatic enzyme induction readily occurs with PBBs, even more powerfully than with PCBs; we do not now know how to measure this effect or appreciate its consequences. Indeed, we are at a loss in relating the extensive observations of in-vivo experiments, with multi-organ dysfunction and damage, to the human condition, which may be subject to different multifactorial interactions, and concomitant or preceding tissue changes of other origin.

Perhaps inadequacies of understanding can be excused by the comparatively recent presentation of chemically induced disease. Pathognomonic syndromes have not been identified. We cannot tell who, exposed to benzidine, will necessarily develop bladder cancer, nor which vinyl chloride polymerization worker will suffer liver fibrosis.

Why chlornaphthalenes produced x-disease in cattle remains a mystery and the mechanisms to explain devastating chick oedema are obscure. Hexachlorobenzene produced death, blindness and porphyria cutanea tarda in Turkish children in 1956; we still do not know the reason.

Peto's stricture may be uncertain in the particular but is surely correct in general. Chemical effects, for the organism, are not all-or-none; one molecule will not upset metabolic balance or the fabric of health. 'Dose' is therefore important, whatever the difficulties of its ascertainment. One day, we shall be able to measure it better. I hope, however, that we shall then also be able to better analyse and quantitate the metabolic and disease response, the health significance of the growing number, amount and complexity of the chemicals in our midst. In this, observations of the effects of inadvertent accidents such as PBB contamination in Michigan can be instructive.

We also share Dr McLean's concern regarding the possibility of reactive depressive illness contributing to our observations. No doubt depression played a role in some cases, as evidenced by 10 % in Michigan reporting they felt depressed compared with 3 % in the Wisconsin farm group reporting similar feelings. Anticipating this complication, Professor Sidney Diamond, neurologist of our examining group, evaluated those felt to have current neurological defects. His clinical judgement was that while in some cases symptoms were definitely depressive, the majority of reported symptoms he evaluated could not be explained in this manner. Further suggesting a toxic syndrome was the observation of the same prevalences and types of symptoms among consumer groups that had purchased dairy farm products from the affected farms but who had undergone no economic or livelihood loss, had not been embroiled in political or legal action, and were unaware of the extent of their exposure. Symptom prevalences and PBB levels were identical in the consumer and farm groups, i.e. those who had run the farms and those who had only eaten produce from them. Before our survey there was no prior evidence to suggest that consumers of non-quarantined products were at risk. Yet the prevalence of symptoms and laboratory abnormalities were the same.

Our examinations came 3 years after the initial episode and thus our observations were of chronic rather than acute effects. This fact is borne out by the calculation of incidence rates for the appearance of new symptoms. For most symptoms, rates peaked in 1974–5 and by 1976 incidence was only slightly higher than for before 1973. We have no measure of blood or tissue levels in 1974 or 1975 before blood–tissue equilibrium was achieved. For most agents, current measurements often do not correlate with the presence of chronic symptoms or disease (benzene blood level and aplastic anaemia; drug, solvent or anaesthetic level and hepatitis, vinyl chloride tissue level and liver fibrosis). In fact, it is the rare agent that can be detected at all 2 or 3 years after exposure, although permanent residue may be observed.

Are we measuring the right material? We are reminded of the Yusho (PCB) incident where PCB levels for the most part correlated with severity of disease. Now we are learning that the disease was not necessarily caused by PCBs but rather by a dibenzofuran contaminant. Although the dose–disease relation with PCB levels appeared, it may have been more fortuitous than causal. In our studies of workers only exposed to PCBs, considerably higher PCB burdens than those in Yusho were found, but with no comparable disease. With PBB, we may possibly also be measuring the wrong material. The 'Firemaster' which was fed to livestock was not uniform and human exposure occurred after possible metabolism and mediation by the animal, adding further confounding effects. The gas chromatographic tracings of serum from directly exposed workers are considerably different from those of the farmers, as well as from the original 'Firemaster' compound.

Further, it is of interest that significant objective abnormalities were present and did correlate with symptoms. The high prevalence of abnormal liver function tests and abnormal lymphocyte functions can not be readily explained without considering a toxic cause. Farmers with symptoms of excessive fatigue had a prevalence of elevated serum glutamate–pyruvate transaminase (SGPT) values, twice that of those without this symptom (18 % as against 7 %). Similar associations with SGPT were observed for the overall group of 'neurological' symptoms as well as for joint pain. These clinical findings speak to the 'realness' of the symptoms reported.

Proc. R. Soc. Lond. B. **205**, 157–164 (1979)

Printed in Great Britain

Occupational experience

By K. P. Duncan†

Health and Safety Executive, Baynards House, 1 Chepstow Place,
Westbourne Grove, London W2 4TF, U.K.

The use of occupational information to find out about causation of disease is a practice of long standing. The working population is partly selected and partly self-selected. It cannot fully represent the general population since it excludes children, some disabled people or those susceptible to disease, and some women. Exposure to a contaminant at work in most cases is limited to 8 h a day. However, in many cases the exposures are much higher than those in the outside environment and, if total dose is the main damaging factor, the hazard is consequently higher. For this reason the first place to seek information on environmental impact will often be the workplace.

Even if the hazard is not large, correlation may be possible if the associated lesion is unusual. Most difficulties occur when a relatively common condition is involved, and when exposures to the possible agent are so low that the population required to show significant statistical variation will be unreasonably high. In this situation, individual attribution to cause may be unrealistic against the background of naturally occurring conditions. Earlier detection by biological indicators may be suggestive but too non-specific to be decisive.

Additionally, secondary contamination (neighbourhood cases) may by its geographical distribution enable hazard identification to be made and dose–response estimated.

The risks associated with new substances have led to the introduction of proposals for a notification scheme. How this is to be related to specific follow-up and analysis is the big epidemiological problem of the time. Mortality studies are too slow, though large-scale ones provide an ultimate test of any theory of dose and response.

This paper is concerned with prospects for the future, and past events are not considered except where they provide examples of value for future thinking. It presents essentially a practical viewpoint, looking at environmental problems at large from the point of view of occupational health practice and including some comments about the relevance of the Health and Safety at Work etc. Act 1974. The unifying theme is the relevance of occupational exposures to various agents and the use that can be made of this type of experience in broader studies. Only long-term effects are considered.

† Present address: Health and Safety Executive, 25 Chapel Street, London NW1 5DT, U.K.

RESERVATIONS

The occupationally exposed population varies from the general population in several obvious ways. Children are excluded, women are still under-represented and the sick or handicapped are variously selected. There is a sizeable problem caused by the fact that considerable self-selection occurs; an occupational group may therefore understate the degree of potential effect. The classical case to illustrate this point is that of exposure to coal tar where 'tar smarts' occur in a proportion of those first exposed, particularly in sunny conditions, and this leads to many such men leaving the relevant industry. It might well be that that particular group would be more sensitive to long-term effects as well, but the selection is made before that point could be determined.

Perhaps the greatest reservation concerns the practical difficulties involved in obtaining accurate and representative measurement, not of the atmosphere in general, but of the probable uptake of a material. This is often overlooked by those not personally familiar with the science of occupational hygiene. The problem is not usually one of the design of instruments, nor of their calibration nor of the development of analytical techniques, but of the microclimate created by the habits of work and behaviour of the exposed workman. There have been frequent occasions where careful measurements and analyses to several places of decimals have been carried out, but the end result may misrepresent the true intakes by factors of 10, 100 or more. This feature is of great importance when the further use of these data particularly in the assessment of dose–response is reached. It is not a serious overstatement to say that all descriptions of exposure should be looked at very carefully, all general statements of exposure of groups are even more suspect, and all the resulting multiplications should be viewed with the greatest suspicion.

Biological monitoring, while it has the obvious advantage of being directly related to the individual concerned, contains its own problems when attempts are made to use it for dose or body burden assessment. The time of assimilation of the material, individual variations in dietary and other habits, and interaction of host and contaminant, add up to formidable difficulties which call for careful scientific appreciation. There are many uncertainties associated with this type of work.

The importance of the completeness of the study population can hardly be overestimated. It is a frequent experience in occupational health work to be informed that no hazard of some contact or exposure occurs and to have this statement supported by evidence of long survival and good health among certain conspicuous members of the exposed group. This type of anecdotal support is not really useful as negative evidence for the various self-selection factors which have already been discussed. Nevertheless, there is no doubt that properly established negative results are of great value.

Assumptions

Occupational intakes will often be uneven. They will in general be over a 48 h week with periods of 'recovery'. They will usually be very much higher than anything that would be expected to occur among the general population, unless under disaster conditions which are not being considered here. In the more normal case of intermittent or chronic exposures, it is essential, if medical or other scientific information arising from studies of these exposures is to be used, that certain assumptions relating the effects of these relatively high doses and the effects to be predicted at lower doses must be made. These assumptions bring in a great many biological considerations, though for protection purposes, especially when talking in terms of carcinogenesis, the no-threshold linear relation has become fashionable in the field of standard setting and would on the face of it appear if it errs at all to err on the side of safety. There are difficulties which have to be allowed for here; statements about dose–response should bear these considerations in mind and not tend to overstate the certainty of their construction. For example, variations in the length of the latent period, considerations of basic biological mechanisms, thoughts about possible recovery processes, and thoughts about synergistic action of different materials or factors of individual diathesis, may make some of these assumptions very complex and dubious indeed. The limitations which should be applied to this type of extrapolation are well summarized in the recent Recommendations of the International Commission on Radiological Protection (1977). Furthermore, there is the problem of deciding the prognostic significance of some of the very early biological effects which may be visible and deciding the likelihood or inevitability of subsequent serious and significant consequential damage.

All of these biological and toxicological events are taking place against a background of natural ageing and of exposure to other possibly damaging constituents in the course of an individual's life. Allowance has to be made for these factors also in determining the special significance of particular exposures to foreign materials. Not infrequently it is possible to be fairly certain that a substance is potentially damaging without having any precise idea of the extent of severity of its possible threat.

Straightforward examples

Tar and substances derived from tar provide possibly some of the best examples of both the successes and difficulties of working from an industrial experience to general statements about health effects. The classic description of soot cancers given by Percival Pott (1775), the subsequent work which led to the identification of the active agent (Yamagiwa & Ichikawa 1916; Kennaway 1955), and later work on products of the combustion of coal and exposures to benzpyrene in coal gas works (Doll *et al.* 1965) or in coke oven plants (Lloyd & Ciocco 1969), make it clear that there is a sizeable cancer risk associated with substantial exposures to benzene solubles arising from tar. These were, however, cases where the incidence of the particular disease concerned, cancer of the scrotum, cancer of the lung and cancer of the bladder, was much higher than would be found in any non-exposed group and therefore the association was not too difficult to establish. Similar results have been shown in the dye industry with its related bladder cancers (Tsuchiya *et al.* 1975) and in the more recent example of the occurrence of chloracne (May 1973) after appropriate exposures. The more serious difficulty comes when an attempt is made to associate much lower exposures even to these known hazardous substances to a smaller number of the industrial population where the excess is not statistically visible, though there may be some small excess which is lost in the background of disease in general. All that one can say confidently is that at the high levels disease occurs and it seems scientifically logical to assume that some degree of hazard is associated with lower exposures. It is difficult to see how that hazard is to be properly quantified with limited populations for study, particularly as in the lung cancer case where another main cause in cigarette smoking exists.

In one case the situation is easier. Even if the hazard is not large, correlation may be possible if the associated lesion is unusual as in the case of the occurrence of angiosarcoma of the liver in association with large exposures to vinyl chloride (Selikoff & Hammond (eds) 1975), or the occurrence of nasal sinus cancers related to certain wood dusts (Acheson 1976). This type of identification can be made much more positive if there are additional geographical factors in support to illustrate the relation of cases. The classic example of this is the neighbourhood cases in the beryllium incidences in the 1940s and 1950s (Reeves 1976). Even when all the favourable factors exist there are pitfalls because, if very low exposures are then to be considered, even on a non-threshold basis, a very low incidence of disease would be expected so that the chance of getting significant figures from either a small slightly exposed industrial population or a larger environmental population, even less exposed, is relatively slight. To some extent, how much this matters depends on whether the intention is to correlate cause and effects in an individual case or to indict a substance significantly as being associated with certain biological effects or to fix allowable concentrations for exposure of any group or for everyone to a particular substance. It is extremely important to separate these various purposes and not to slip by oversimplification from one to the other. There is a big difference between the study of a problem to ascertain a scientific association and the decisions that must be taken about actions that should follow.

Many of the difficulties that have been considered in this section have been well illustrated by the study of Hems (1966) who considered what population and what radiation exposures would be needed to show up, on a non-threshold assumption, a significant increase of leukaemia. It has to be recognized that just as absolute safety is an unobtainable goal, absolute scientific certainty of association, particularly in individual cases or small numbers, is also a very uncertain thing.

DIFFICULTIES

If it is sometimes possible, for certain purposes at any rate, to discuss cause and effect relations between rather unusual diseases which occur frequently and some environmental exposure, or between very dramatic events and consequent effects, then it also has to be faced that there are particular difficulties which occur when the event which is being studied is a common one and the distance in time between the original exposure and the biological event is long. It is perhaps unnecessary crystal gazing but it does seem likely that many of the worries that will arise in the future will concern not so much obvious, distinct and easily attributable occupational or exposure effect but rather a raised occurrence of relatively common conditions. Obvious examples, other than those in the field of carcinogenesis which are those usually quoted, are cardiovascular effects, chronic neurological effects and the common diseases like chronic bronchitis and high blood pressure. In these cases it seems inevitable that there will have to be particular studies aimed at selected groups rather than an assumption that any real hope exists of finding by a general system some relatively small, at least in the differential numerate sense, effect. At the same time the increasing degree of public awareness and public concern means that we have to consider the best things that we can do to avoid the long delay which in the past has appeared almost inevitable. This is further compounded by the fact that people change their jobs frequently and by the fact that, even within the one factory or the one firm, a man may have multiple exposure to many different agents in one year, never mind throughout his life. To put it at its simplest, it is not very difficult, or at least it is within quite reasonable bounds of expectation, to produce a register of everybody who has been exposed to substance A and to have these people marked in some way for future morbidity and mortality studies but it is very difficult to see how one can deal with the fact that they have also been exposed to substances B, C, D, etc., in the interim. It may be that they will appear on separate registers for these substances as well, in which case the allocation of responsibility for the effect is going to be very difficult but it may very well be that these substances are in some degree synergistic or that they have had an as yet undetected effect so that a false attribution to substance A becomes rather likely.

FUTURE NEEDS

In spite of all the difficulties it is perfectly obvious that some improvement in the present system is absolutely essential. Towards that end, the Health and Safety Commission has recently issued a Consultation Document about the Notification of New Substances (1977). This does mean that at least there will be a register of new substances being brought into use and that certain basic data about these substances will be recorded. The extent of recording is not yet determined, but it must be made clear that this is a register and not some sort of certificate of clearance. Very recently a Discussion Document on Occupational Health Services (1977) has been published in which the need for much better employment records to be maintained is emphasized and industry itself is urged to consider the construction of suitable population studies. It should be emphasized that the Health and Safety at Work etc. Act 1974 places the responsibility for ensuring the safety of the introduction of substances, their description and their operation plainly upon the manufacturer or the employer concerned. All sorts of problems are raised here and some of these are discussed elsewhere in this symposium. A significant point is that much of this testing and recording work will have to be done by industry itself; it is therefore important that the public, employed population, government, whoever else is concerned, is satisfied of the scientific integrity of existing occupational health services. It is essential to point out that general suspicion cast on those already working in the industrial field is very counterproductive; many of the best skills and the greatest body of knowledge arises among these people – scientists, doctors, administrators and others. What is quite clear is the message of the Robens Report (1972) and of the Health and Safety at Work Act, that it is absolutely essential for there to be full consultation on all these matters and for these actions to be taken openly and with total scientific integrity. Some of the dangers of fashionable sniping at established occupational health services are discussed elsewhere in more detail (Duncan 1977).

The future need is for the construction of some system of record keeping that goes in parallel with other moves that may be taken about notification of substances and ensures that as far as possible, the working population where the first warning signs of any ill-effect will probably occur, can be observed. As a practical contribution towards that, it is very possible that we ought to be looking at the introduction of some sort of early warning system or notification of adverse effects from industrial doctors, other safety professionals, from Health and Executive staff and, as far as this may be practicable, from general practitioners, but the details of any such recording or alerting scheme will take a good deal of working out. Probably we should combine simple general schemes with scientific closely pointed approaches. The price of liberty is eternal vigilance and that may also be the price of advancing technology.

Conclusion

The practical difficulties are certainly large. There is, however, no need to think that a perfect solution is the only one worth seeking and the important thing perhaps is to introduce stage-by-stage steps towards an improvement in the present rather haphazard situation. The very difficulties of the problem and the attention given to it are a measure of the improvements which have been brought in since the pace of technology increasd. They are a measure also of our unwillingness to accept excessive risks and of the need to reduce the hazard. In our desire for perfect scientific solutions perhaps we should not too readily underestimate the value of empirical decisions and the value of general statements about reducing exposures to the lowest practicable level. Not all of these reductions and not all of these decisions are taken on purely scientific grounds, but that does not in any way invalidate the contribution to the increased health and safety of a workforce and it does not help for too sophisticated a view to be formed or for scorn to be poured on these somewhat *ad hoc* practical actions until such time as truer scientific evidence can be produced. Occupational studies and occupational experience are essential as the first step towards detecting effects and one of the aims of this paper is to introduce a note of perhaps rather unscientific caution into the present debate on the most effective ways of producing improvements. At the end of the day anything that reduces unnecessary ill-health is good, provided it can be sensibly introduced within a practical framework, unless we are to take the view that those who persecuted Galileo were right and technological advance should be prohibited.

Attitudes are changing. There is a great future in the tripartite open approach involving cooperation between managements, unions and government. More detail is spelt out in the Health and Safety Commission's Discussion Document on Occupational Health Services where it is recognized that there are practical problems but that they can be solved. There are attitude problems of emotionalism and suspicion and these must first be resolved.

There is a Latin tag 'Ubi emolumentum ibi onus' which is often quoted as referring to management responsibility. That must be a shallow interpretation in this generation. The rewards are for management and employed alike and the third force is society in general as represented by government. If the right path can be chosen in industry, that will establish the fullest front-line contribution to the detection of long-term environmental hazards to health. It will not be easy, but that is the contribution that occupational health can and must make.

REFERENCES (Duncan)

Acheson, E. D. 1976 Nasal cancer in the furniture and boot and shoe manufacturing industries. *Prev. Med.* **5**, 295–315.

Discussion Document on Proposed Scheme for the Notification of the Toxic Properties of Substances 1977 Health and Safety Commission. London: H.M.S.O.

Doll, R., Fisher, R. E. W., Gammen, E. J., Hughes, G. O., Tyrer, F. H. & Wilson, W. 1965 Mortality of gasworkers with special reference to cancers of the lung and bladder, chronic bronchitis and pneumoconiosis. *Br. J. ind. Med.* **22**, 1–12.

Duncan, K. P. 1977 Future developments in occupational health. *J. R. Soc. Arts* **125**, 684–689.

Health and Safety at Work etc. Act 1974 Chapter 37. London: H.M.S.O.

Hems, G. 1966 Detection of effects of ionising radiations. b. Population studies. *Br. med. J.* i, 393–396.

Kennaway, E. L. 1955 The identification of a carcinogenic compound in coal tar. *Br. med. J.* ii, 749–752.

Lloyd, J. W. & Ciocco, A. 1969 Long term mortality study of steelworkers. I. Methodology *J. occup. Med.* **11**, 299–310.

May, G. 1973 Chloracne from the accidental production of tetrachlorodibenzo-dioxin. *Br. J. ind. Med.* **30**, 276–283.

Occupational health services: the way ahead 1977 Health and Safety Executive. London: H.M.S.O.

Pott, P. 1775 *Chircurgical observations*. London: Hawes, Clark & Collings.

Recommendations of the International Commission on Radiological Protection 1977 I.C.R.P. Publications 26. Oxford: Pergamon Press.

Reeves, A. L. 1976 Berylliosis as an auto immune disorder. *Ann. clin. Lab. Sci.* **6**, 256–262.

Safety and Health at Work 1972 Report of Robens Committee 1970–72. Cmnd 5034. London: H.M.S.O.

Selikoff, I. J. & Hammond, E. C. 1975 Toxicology of vinyl chloride – polyvinyl chloride. *Ann. N.Y. Acad. Sci.* **246** (31).

Tsuchiya, K., Okubo, T. & Ischizu, S. 1975 An epidemiological study of occupational bladder tumours in the dye industry of Japan. *Br. J. ind. Med.* **32**, 203–209.

Yamagiwa, K. & Ichikawa, K. 1916 *Experimentelle Studie uber die Pathogenese der Epithelialgeschwulste*. Tokyo: Tokyo University Faculty of Medicine.

Proc. R. Soc. Lond. B. **205**, 165–178 (1979)

Printed in Great Britain

Record linkage and the identification of long-term environmental hazards

By E. D. Acheson

University of Southampton Medical School, South Block, Southampton General Hospital, Tremona Road, Southampton SO9 4XY, U.K.

The long-term effects of toxic substances in man that have been discovered so far have involved gross relative risks or bizarre effects, or have been stumbled upon by chance or because of special circumstances. These facts and some recent epidemiological evidence together suggest that a systematic approach with better methods would reveal the effects of many more toxic substances, particularly in manufacturing industry.

Record linkage is a powerful tool because it makes possible the correlation of indicators of exposure with indicators of the biological effect of such exposure in the same persons or in their progeny even after considerable periods of time have elapsed. A system of linked records exists in England and Wales which is at present used by research workers to follow up samples of persons defined in various ways, e.g. in respect of exposure to a suspected toxic factor. In this way hypotheses about substances causing cancer or other lethal effects can be tested.

It is suggested that there are two additional ways in which record linkage techniques could be used to identify substances with long-term toxic effects: the first would be by setting up a register of women employed in industry during pregnancy and linking this register to records of the occurrence of congenital malformations and of stillbirth or death in their children; the second would be to follow samples of workers in manufacturing industry, notably those engaged in the manufacture of products from raw materials including the chemical industry, to death and to the development of cancer. Regular analyses of material from these two systems of linked records would provide the basis for a monitoring system for certain gross effects of long-term toxic substances in man.

There are two principal obstacles to further progress in this field. The first is the lack of a clear statement of public policy concerning the issues of confidentiality and informed consent in the use of identifiable personal records for medical research. A settlement is needed which defines the proper limits of their use in the interests of health with safeguards to privacy. The second obstacle is a lack of resources to improve the quality, accessibility and organization of the appropriate data.

1. The problem

The toxic factor to be identified may be any substance introduced by man into his environment as a result of the application of science and technology in industrial processes, including the processing and storage of food and water. For the purpose of this paper the long-term effects of physiological and psychological environmental factors such as inadequate or excessive exertion, faulty working posture and external trauma and stress are excluded, as are the long-term effects of drugs and of substances like tobacco, which are taken voluntarily. Occupational exposure is of particular interest because higher concentrations of the substances are usually present in the workplace than in the general environment, and prolonged contact between the substance and a restricted group of persons improves the chances of detection of a given biological effect. Occasionally, however, an industrial effluent, or interaction between an effluent and an external factor may place a population outside the workplace (e.g. in the vicinity of a factory) at greater risk than the workforce itself.

It is obvious that there can at present be no reliable estimate of the total burden of sickness and mortality due to long-term toxic effects as defined above. The general continuing decline in mortality in most age groups in such highly industrialized countries as Sweden, the Netherlands and England and Wales suggests that up to the present the combined effect of such factors has not offset the beneficial effects of better housing, hygiene and nutrition. Up to the present (and one should remember that at present we are witnessing the biological effects of factors introduced into the environment say 20 years ago) the total burden of toxin-induced sickness is probably much smaller than that due to the physiological and psychological changes in environment and habit mentioned above.

The definition of 'long-term' is arbitrary. We are concerned with factors which produce a biological effect in a person after a delay sufficiently long to make it difficult to associate the effect with its cause. Occasionally the effect may be seen not in the person exposed but in the immediate progeny (as in the case of a teratogen acting upon women in early pregnancy) or, in the case of a mutagen, in other descendants of the exposed person. For known toxic factors there is a spectrum of time intervals between first exposure and the appearance of the biological effect. Teratogens operating in early pregnancy with effects obvious at birth (*ca.* 250 days) are examples at one end of the distribution. At the other extreme are leukaemia and solid tumours following exposure to ionizing irradiation (*ca.* 20 years), tumours following exposure to polycyclic hydrocarbons, dyestuffs and asbestos (20–60 years), and the operation of mutagens. Two important practical implications emerge from these long intervals. The first is that, owing to mobility of labour and migration, the environmental circumstances associated with the cause are unlikely to be evident when the effect manifests itself. The second is that the preservation and use of data about past exposure will be an essential part of any monitoring system if questions posed are to be answered within a reasonable span of time.

2. THE DISCOVERY OF LONG-TERM TOXIC EFFECTS UP TO THE PRESENT

Before attempting to assess how record linkage techniques may be used to identify long-term hazards from man-made chemicals, it may be helpful to examine the ways in which such substances have been identified up to the present. By far the most usual method has been the association of cause and effect by a clinician or pathologist studying sick people or tissues derived from them. An extraordinary concentration of cases is noted by an observer and related to an environmental factor shared by the patients concerned, usually associated with an occupation. Examples taken from the field of carcinogenesis are shown as table 1, but similar arguments apply to other fields, for example the neurotoxins.

TABLE 1. OCCUPATIONAL CARCINOGENS BY SITE, OCCUPATION, AND METHOD
BY WHICH THEY WERE FIRST DISCLOSED (AFTER DOLL 1975)

(a) *First noted by clinicians or pathologists*

scrotum	polycyclic hydrocarbons	sweeps
pleura	asbestos	miners
skin	ionizing radiation	radiologists
	ultraviolet light	outdoor workers
	arsenic	sheep dip makers
nose	wood dust	furniture makers
	isopropyl oil	chemical workers
nose and bronchus	nickel oxide	smelters
leukaemia	benzene	leather workers
bone	radium	luminizers
bladder	2-naphthylamine	dye workers
bronchus	chrome pigments	refiners
	bichloromethyl ether	ion exchange makers
liver	vinyl chloride	PVC manufacturers

(b) *First noted in work in animals*

bronchus, nose, larynx	mustard gas	chemical workers
bladder	4-amino diphenyl	chemical workers

(c) *First noted by epidemiologists*

bladder	2-naphthylamine	coal gas producers
	2-naphthylamine	rubber workers
nose	leather dust	boot and shoe operatives

Special situations, favourable to the discovery of the factor concerned, have been operative in many of the cases quoted. In some instances, e.g., hepatic angiosarcoma, nasal cancer, bone sarcoma and mesothelioma, the natural frequency of the tumour has been so low that the occurrence of a handful of cases in the experience of one group of observers in one environment was sufficient to attract attention. In others, e.g. the concentration of 3000 furniture workers in the small town of High Wycombe or of 30000 boot and shoe operatives in Northamptonshire, the topography of the industry has been a major factor in bringing the cases of the disease in question to the attention of a group of observers, or a bizarre distribution of the tumours (as in arsenic induced skin cancer) or their association in the same person with other effects of the agent (e.g. lung cancer and asbestosis) have been instrumental in the discovery. In one study the coincidental existence of a large rubber factory in the same area as one of the dyestuff factories which were the subject of the original enquiry was the factor which determined the discovery that there was also a risk associated with the use of 2-naphthylamine in that industry (Case & Hosker 1954).

At least as far as carcinogens are concerned, the demonstration of an effect in laboratory animals has so far rarely been the initiating factor in the discovery of an effect in man. Doll found only four examples: carcinoma of the bronchus in gas retort house workers, due to polycyclic hydrocarbons; respiratory cancer in mustard gas manufacturers; angiosarcoma of the liver in manufacturers of polyvinyl chloride; and bladder cancer in chemical workers using 4-amino-diphenyl (Doll 1975). Similarly, although they have almost always been necessary for further elucidation and proof, studies with the use of epidemiological techniques have only occasionally brought the existence of a toxic factor to first notice. It is worth noting that, with the possible exception of alcohol, the action of none of the known carcinogens was first detected in the Occupational Mortality tables published every 10 years by the Office of Population Censuses and Surveys (1861–1961). This is the only systematic material monitoring the effects of the industrial environment upon health at present.

As there is no logical reason why long-term toxic effects should manifest themselves predominantly in special situations or in terms of tumours which are usually rare, or by means of other bizarre effects, it is safe to assume that many substances that have induced increases in sickness and mortality have so far gone undetected. Up to the present we have detected only those factors that reveal themselves because the effect is gross or where we have stumbled upon them with the help of luck. We have had no systematic approach to the problem and few effective tools. The situation might be compared with that in astronomy before the invention of the telescope, or in biology before the optical microscope. A second line of evidence that suggests that many carcinogens exist undetected has been provided by some recently published studies of the relation of the occurrence of cancer in the United States to the distribution of the chemical and metal refining industries. Hoover & Fraumeni (1975) found, for example, relations between high mortality from bladder cancer in certain U.S. counties and the manufacture in those counties of dyes, pharmaceuticals, cosmetics and printing ink. These and some other associations demonstrated by the same workers are summarized in tables 2 and 3. In another study, Blot & Fraumeni (1975) found excesses of lung cancer in U.S. counties with copper, lead or zinc smelting and refining industries but not in counties where non-ferrous ores are processed. Although such studies can never provide conclusive proof of causal relations, when sufficiently plausible, as in this case, they point to an urgent need for comprehensive data about the workforces in the industries concerned.

A further lesson which has practical implications for the future, which may be learnt from past experience, is that if drugs and toxic substances taken socially are excluded, the vast majority of long-term toxins so far found in man have been identified in manufacturing industry including mining and refining. As far as

TABLE 2. CORRELATIONS BETWEEN TYPES OF SUBSTANCE MANUFACTURED AND EXCESS CANCER MORTALITY IN 137 U.S. COUNTIES REPORTING ANY CHEMICAL INDUSTRY (HOOVER & FRAUMENI 1975)

bladder	lung	liver
dyes, pharmaceuticals, cosmetics, printing ink	gases, pharmaceuticals, soaps, paints, synthetic rubbers	organic chemicals, rubber, soaps, cosmetics, printing ink

TABLE 3. CANCER MORTALITY IN 137 U.S. COUNTIES REPORTING ANY CHEMICAL INDUSTRY: SITES OTHER THAN BLADDER, LUNG AND LIVER WITH EXCESSES OVER 10% ABOVE U.S. NORM AND SIGNIFICANT AT $p < 0.05$ (HOOVER & FRAUMENI 1975)

men	women
nasal sinuses, larynx, skin, bone, melanoma, mouth, throat	nasal sinuses, nasopharynx, melanoma, throat

carcinogens are concerned, the point can be made by an analysis of a comprehensive list of occupations in which carcinogenic factors have been identified (Cole & Goldman 1975). Of 55 occupations listed, 39 (71%) were in the manufacturing sector of industry, 8 (15%) in the mining and production sector and 7 (13%) in the transport distribution and service industries.

3. RECORD LINKAGE

Record linkage is the bringing together into a single personal file of information recorded at different times or places about an individual. In the context of this paper, record linkage is of interest in that it facilitates the bringing together of information about *exposure* of members of a population to an agent, with information about the *occurrence* of a disease in that population, even when such events are separated by substantial distances in space and time. In order to link the information it is necessary that each record should contain enough information about the person's identity to enable his record to be distinguished from those of all others in the population. Although it is highly desirable that an identification cipher unique to the individual should be present on as many of the relevant records as possible, computer techniques of identification are available which can utilize a wide range of information including names and date of birth. These techniques can accomplish linkages of machine readable records on a large scale cheaply and more effectively than can clerks (table 4). If linkage of records is to be carried out effectively on a large scale it is also necessary to ensure that the files to be linked are organized and sequenced according to certain rules.

Record linkage may be *ad hoc* in the execution of a particular study, e.g. Case's linkage of a register of employees in the dyestuffs industry with a register of deaths from bladder cancer (Case *et al.* 1954). Cancer registers and unit medical

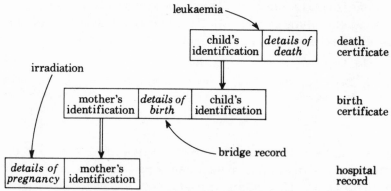

FIGURE 1. Use of a bridge record to link records of two generations.

TABLE 4. COMPARISON OF THE EFFECTIVENESS OF THE LINKAGE OF 44 000 RECORDS OF ILL-HEALTH IN CHILDREN TO THEIR BIRTH REGISTRATION RECORDS, MANUALLY AND BY COMPUTER

(British Columbia data (Smith, M. & Newcombe, H. B., to be published).)

method of linkage	missed linkages (%)	false linkages (%)
computer	1.5	0.08
manual	3.0	2.3

record systems employ continuing record linkage on a limited basis in order to avoid duplication of their records due to multiple notification of the same case. Ongoing systems of linked records, which incorporate some of the medical records of defined populations into a series of personal cumulative files, have also been set up (Acheson 1967). Family record linkage brings together the experience of two or more relatives by the use of a *bridge record*, i.e. a record (e.g. a birth or marriage record) which identifies two or more members of the family (figure 1), enabling, for example, exposure in a mother to be related to a biological effect in a child.

Record linkage techniques may assist in the identification of long-term toxic effects in man in two main ways:

1. The first is by facilitating the testing of hypotheses derived from other types of evidence, e.g. clinical studies, bioassay studies in animals or in-vitro systems. Under this heading record linkage techniques are used to follow cohorts of persons classified in terms of exposure to various factors to the point of occurrence of disease or to death. Substantial facilities for hypothesis testing with the use of

manual record linkage techniques already exist in the U.K. and will be described below.

2. The second way is by systematic ongoing analysis of the comparative risks of disease and death in populations in terms of such indicators as can be made available on a large scale of their exposure to toxic factors. Although such analyses would not provide conclusive evidence of the existence of toxic factors, they might be used to provide a monitoring system which, albeit crude, would be an improvement on the present situation. The indicator of exposure which could be made available most easily would be date, place and duration of employment for a sample of the population employed in a range of processes in manufacturing industries. The World Health Organization, among others, has recently supported the setting up of such systems (W.H.O. 1973, 1974).

4. Records currently available for linkage

Records relevant to the subject of this paper may be divided into two classes: those containing information about sickness and death, and those from which inferences may be made about exposure of a person to a particular substance. In both classes, if linkage is to be achieved, each record must include sufficient particulars of identification of the person concerned and the files of records must be organized in a way that permits the appropriate comparisons between possible pairs of records to be made rapidly.

Morbidity and mortality data that meet these requirements in England and Wales are limited to death certificates and cancer registrations, the usefulness of the latter being limited to some extent by unevenness of ascertainment. Records of congenital malformations detectable at birth have been collected by the Office of Population Censuses and Surveys (O.P.C.S.) on a national basis for some years, and regular computer searches of this material are made for significant clusters of specific malformations. Unfortunately, although it is possible for the mother's identity to be established and her hospital records to be traced on an *ad hoc* basis via the local community physician, the reverse process, which would permit linkage with data about exposure, namely the identification of mal-formed babies from a sample of mothers, is not possible. In Scotland, but not in England and Wales, information is available for linkage in respect of all illness treated on an in-patient basis, and for children in respect of school health examinations and the registration of handicap (M. A. Heasman, personal communication).

As far as records relating to exposure of identified persons to supposed toxic substances are concerned, there is, with the exception of the special study involving census data carried out by O.P.C.S. and mentioned below, at present no systematic source of information available for linkage with morbidity data. Evidence recently submitted to the Advisory Committee on Asbestos showed that many firms maintain fully identifiable records of personnel for periods of up to 40 years which at least establish the presence and duration of employment of a given worker in a particular factory, and on some occasions could be used to give more precise data about the substances to which the workers were exposed (Health and Safety Executive 1977). Although such data give only a first approximation of the dose of a substance received by the worker they may be sufficient to establish a *prima facie* case of risk in certain circumstances. In other industries, for example the chemical and oil industries, more detailed information about exposure is known to exist and might be used to pinpoint a particular risk suggested by a general monitoring system. As there are at present no legal requirements for retaining these types of data, much no doubt has been and will continue to be destroyed.

5. FACILITIES FOR TESTING HYPOTHESES BY USING RECORD LINKAGE TECHNIQUES

The National Health Service Register and Index at Southport contains a record of the identity and postal address of every person registered with a doctor in England and Wales. This register is updated for births and immigrations and cleared of deaths and emigrations by means of a large-scale manual operation. Similar registers exist for Scotland and Northern Ireland. Cohorts of identifiable persons defined for medical research purposes may be followed through this register for death, emigration and the development of cancer. Up to the beginning of 1976, a total of 284 samples of persons had been or were in the process of being followed to death or the development of cancer through this system. Although a number of these studies were very small, a few involved many thousands of people. Some of these studies are concerned with the detection or measurement of long-term toxic effects (e.g. follow-up of vinyl chloride workers); others are designed to study the prognosis of disease. A weakness of the system as it stands today is that *it provides no information about non-fatal illness other than cancer*. The gravity or otherwise of this omission depends upon the likelihood of long-term toxic chemicals having important effects on health other than the genesis of cancer and lethal diseases. Mental and rheumatic diseases are possible examples.

In Scotland a system exists which permits follow-up of defined groups of exposed persons in respect of *all illness treated on an in-patient basis* (including cancer and mental illness). It is also possible in Scotland to follow cohorts to death and to follow children in respect of school health examinations and the registration of handicap.

The addition in 1969 of key items of identification of the deceased person to the death certificate in England and Wales (maiden name, birth place and date of birth) has facilitated record linkage for death certificates. For example, it is now possible to follow women identified in terms of name *before* marriage (e.g. where they had worked in an industrial environment before marriage) to death where this takes place after marriage. However, because there are inadequate identification data relating to the mother on the record it is impossible at present to use the O.P.C.S. file of congenital malformations detected at birth to determine the outcome of pregnancy in women exposed to chemicals in industry. As none of the facilities mentioned above provide regular systematic analyses of biological effects in terms of exposure, they cannot be said to fulfil a monitoring function.

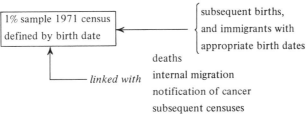

FIGURE 2. The O.P.C.S., standing 1 % sample of the population of England and Wales. (O.P.C.S. 1973.)

6. RECORD LINKAGE AND MONITORING THE ENVIRONMENT FOR TOXIC CHEMICALS

The use of the occurrence of human disease as the end point in monitoring the environment for toxic chemicals has the obvious disadvantage that by the time the effect is manifest the substance concerned may be widely disseminated. Nevertheless, experience with animal and in-vitro bioassay systems suggests that for the foreseeable future the collection of information derived from man will continue to be important in the identification of toxic factors.

Since 1971 a 1 % sample of the population of England and Wales (i.e. about 500 000 persons), kept up to date for subsequent births, deaths, immigrations and emigrations, has been linked to occupational and other personal data recorded about these people at the census and to the National Cancer Register (figure 2) (O.P.C.S. 1973). The linkage procedures used protect confidentiality by being anonymous and by being carried out within the O.P.C.S. with strict safeguards. As the data from successive censuses are added to the file it will be possible to define samples of men and women in terms of their occupational or residence experience and make systematic comparisons of their cancer incidence and mortality pattern with that of the whole sample. This is an important initial step in the direction of a monitoring system. Unfortunately, as can be seen from tables 5 and 6, a 1 % sample would contain very small numbers of men and women from those industries where exposure to toxic chemicals is most likely to take place, and little could be expected from analysis of the experience of the sample as it is at present selected.

In Canada the cumulative employment experience, derived from annual returns of occupation and industry, of 10 % of the individuals contributing to Unemployment Insurance during the period 1942–71 have been linked to mortality data over a corresponding period, and the results of analysis will shortly become available. Within the limitations imposed by small numbers it will be possible to compare the risks of mortality from common conditions of persons employed in various industries with standards derived from the experience of the whole sample (Newcombe 1974).

Monitoring systems of the health of employees are being set up by certain large firms within industry. These have the potential of providing much more detailed information about the nature and duration of exposure of personnel to substances than official data recorded at the census or for administrative purposes. An important drawback, however, is that they may have no access to information about personnel who have left the firm concerned. Such infromation is crucial in the identification of risks of disease with long latent periods.

7. SUGGESTIONS FOR THE FUTURE

A number of important advances in the organization of medical information which have occurred in the last 10 years and which have been mentioned above have improved our capacity to identify and measure long-term toxic effects. Although some of the suggestions made below would involve expenditure, planning could start immediately in the hope that they might be implemented when the national economic situation is more favourable.

Suggestions to improve ascertainment and organization of information about sickness and death

A review of the organization and degree of ascertainment of the cancer registers should be undertaken. Particular points that require attention are the extent to which the data recorded throughout the U.K. may be simplified and standardized further and the extent to which a single data capture process can cover the needs both of cancer registration and other requirements for morbidity data. There is also an urgent need for review of the current regulations concerning the destruction of medical records in the N.H.S. At present, according to HM(61)73 (National Health Service 1973), medical records and allied documents including clinical notes, blood transfusion records, prescriptions and surgical operation books, may be destroyed *6 years from the date of the last entry*, or, where the patient dies in hospital, *3 years after death*. While it is clearly stated in the regulations that summaries of clinical notes and records of post-mortems should be retained, it is not certain whether this is being complied with. In view of the very long latent periods of action of some toxic substances in man, it is desirable that at least a summary of each in-patient admission and of the events associated with pregnancy and delivery should be retained in the principal centres for a minimum of 20 years. The national file of congenital malformations should be modified to contain sufficient information to identify the mother. This would permit linkage to records of occupational and other exposure in pregnancy. An extension of morbidity data available for linkage in England and Wales to include other types of illness other than cancer treated in hospital (as in Scotland) would be time-consuming to organize and would require careful justification.

TABLE 5. NUMBERS OF EMPLOYEES IN FULL-TIME EMPLOYMENT BY INDUSTRY ORDER IN GREAT BRITAIN, 1975, EXPRESSED IN THOUSANDS

(*British Labour Statistics Year Book 1975* (1977) London: H.M.S.O.)

		employees			
		male		female	
Order†	type of industry	number	(%)	number	(%)
I, II, XX, XXI	production	2031	(16.2)	184	(3.4)
III–XIX	manufacturing	5083	(40.5)	1647	(30.4)
XXII–XXIII	transport and distribution	2258	(18.0)	960	(17.7)
XXIV–XXVII	service	3168	(25.3)	2630	(49.5)
	total	**12540**	**(100.0)**	**5421**	**(100.0)**

† Standard Industrial Classification, 1968.

TABLE 6. NUMBERS OF FULL TIME EMPLOYEES ENGAGED IN THE MANUFACTURE OF PRODUCTS FROM RAW MATERIALS, EXPRESSED IN THOUSANDS

(*British Labour Statistics Year Book 1975* (1977) London: H.M.S.O.)

		employees	
Order	type of industry	male	female
IV	coal and petroleum products	35	4
V	chemical and allied industries	302	96
VI	metal manufacture	441	46
XVI	bricks, pottery, etc.	204	50
XVII	timber, furniture, etc.	204	37
various	others†	130	36
	total	**1316**	**269**

† Makers of flour, tobacco, synthetic fibres, leather and paper.

Suggestions to improve ascertainment and organization of information about exposure to toxic chemicals

A study should be made of the scope, scale and accessibility to data processing machines of information already available about the occupation and industry of identified persons held in the various Departments of Government and in industry. A cumulative occupational record should be maintained for research purposes for a sample of the work force (as in Canada) (Newcombe 1974). The sampling system should take cognisance of the special potential risk of workers within certain manufacturing industries, notably the oil, chemical and metal refining industries. The number of employees in these industries in relation to all full-time employed persons is shown in tables 5 and 6. Consideration should be given to the enlargement of the O.P.C.S. sample to incorporate appropriate fractions of manufacturing industry so that this might be linked to cancer registration and mortality records. A special register should be made of women who have worked in industry during pregnancy. This register, classified according to the nature of their work, should be linked to records of congenital malformations and to stillbirth and mortality records in infancy. If, as is commonly held, carcinogens are also teratogenic, such a register might provide an early warning of the release of potential carcinogens into the environment.

8. Obstacles to further progress
The confidentiality issue

Information, including virtually all data about sickness, arising from contact between doctors and patients, is rightly regarded as confidential. Other important data collected in connection with the registration of birth, stillbirth and death, and at the population census, are protected by stringent legislation from disclosure. In recent years, the development of electronic data processing has intensified public anxiety about disclosure of data to third parties and particularly to Government. Legislation is coming into effect in various countries which controls the conditions under which identified data may be stored, disclosed or linked. If serious constraints are not to be placed upon the methods outlined above for the identification of long-term toxic hazards in the United Kingdom, it is essential that any legislation should take account of the *bona fide* needs of epidemiological and genetic research. There is also a need for better public understanding of the need for this type of work and public recognition and commitment to it by Government. The following are particular areas for anxiety from the point of view of the public interest in respect of health.

1. Will limitations be placed on the utilization of data for purposes other than that for which it was recorded? Such a well intentioned proposal (Committee on Privacy 1972), if translated into legislation, might make it difficult or impossible to utilize records of exposure in industry, in epidemiological surveys, unless an exception were made for medical research.

2. Will the informed consent of the patient be necessary before clinical data about him may be used for research? For example, will it be possible to send data about the occurrence of cancer to a cancer register only if the patient knows he has cancer and consents to the transmission of the data? If so, there will inevitably be a serious decline in the notification of cases, and in the value of cancer registers. Research into the existence of possible carcinogens in industry is another area which would be inhibited by such a measure. It would, for example, be entirely inappropriate to seek informed consent from workers to use their clinical records where the strength of the *a priori* evidence of the existence of a carcinogen did not justify raising their fears.

3. Will the further rationalization of ciphers which facilitate record linkage (e.g. N.H.S. and Social Security numbers) and their utilization be discouraged or even prohibited? In the United States, constraints have been placed on the use of the Society Security number for linkage purposes outside the Social Security System. These constraints make epidemiological research more difficult.

The *Report of the Committee on Privacy* devoted a chapter to a sympathetic treatment of the special problems of confidentiality in relation to epidemiological research. It recognized the importance of confidential medical data in epidemiological and genetic research and the need to balance in any legislation the public interest with regard to the promotion of health and the need to protect personal privacy. However, the more recent White Paper *Computers and privacy* (1975) and its supplement *Computers: safeguards for privacy* (1975) give no indication that the special requirements of epidemiological and genetic research are likely to be taken into account in forthcoming legislation.

Lack of resources and political support

The lack of appreciation of the importance of records and their linkage in the identification of the toxic effects of man-made chemicals means that the processing of medical records, the maintenance of disease registers and the training of skilled medical records personnel all have low priority in the competition for resources. In times of economic stringency, as at present, the actual continued existence of the National Cancer Registers in England and Wales and the morbidity data collecting system in Scotland are threatened.

Equally important as a discouragement to further developments in this field is the ambiguity of the Government's approach to medical record linkage. A clear statement of the propriety and importance of this work is urgently needed together with a code of practice to define the proper limits of the use of these techniques in the interests of health with proper safeguards to privacy and legal protection for the custodians of the data.

REFERENCES (Acheson)

Acheson, E. D. 1967 *Medical record linkage.* Published by the Oxford University Press for the Nuffield Provincial Hospitals Trust.

Blot, W. J. & Fraumeni, J. F. 1975 Arsenical air pollution and lung cancer. *Lancet* ii, 142–144.

Case, R. A. M. & Hosker, M. E. 1954 Tumours of the urinary bladder as an occupational disease in the rubber industry in England and Wales. *Br. J. prev. soc. Med.* **8**, 39–50.

Case, R. A. M., Hosker, M. E., McDonald, D. B. & Pearson, J. T. 1954 Tumours of the urinary bladder in workmen engaged in the manufacture and use of certain dyestuff intermediates in the British chemical industry. *Br. J. ind. Med.* **11**, 75–104.

Cole, P. & Goldman, M. B. 1975 In *Persons at high risk of cancer: an approach to cancer aetiology and control.* (ed. J. F. Fraumeni), pp. 167–184. London: Academic Press.

Committee on Privacy 1972 *Report of the Committee on Privacy.* Cmnd 5012. London: H.M.S.O.

Computers and privacy 1975 Cmnd 6353. London: H.M.S.O.

Computers: safeguards for privacy 1975 Cmnd 6354. London: H.M.S.O.

Doll, R. 1975 Pott and the prospects for prevention. *Br. J. Cancer* **32**, 263–272.

Health and Safety Executive 1977 *Selected written evidence submitted to the Advisory Committee on Asbestos, 1976–7.* London: H.M.S.O.

Hoover, R. C. & Fraumeni, Jr, J. F. 1975 Cancer mortality in U.S. counties with chemical industries. *Envir. Res.* **9**, 196–207.

National Health Service 1973 *Preservation and destruction of hospital records.* H.M. (61)73.

Newcombe, H. B. 1974 *A method of monitoring nationally for possible delayed effects of various occupational environments.* National Research Council of Canada N.R.C.C. no. 13686.

Office of Population Censuses and Surveys 1861–1961 *The Registrar General's decennial supplements. England and Wales. Occupational mortality tables.* London: H.M.S.O.

Office of Population Censuses and Surveys 1973 *Cohort studies and new developments.* London: H.M.S.O.

World Health Organization 1973 *Environmental and Health Monitoring in Occupational Health.* (Technical Report Series no. 535.) Geneva: W.H.O.

World Health Organization 1974 *W.H.O. Environmental Health Monitoring Programme. Report of a W.H.O. meeting.* EHE/75.1. Geneva: W.H.O.

Discussion

R. Peto (*Radcliffe Infirmary, Oxford, U.K.*). It would be most valuable if the occupation recorded on all decennial census forms before death could be linked to the death certificate. This would be achieved if photocopies of the census returns were inserted into the record for each patient at the N.H.S. central register at Southport. The cost would be approximately £1 per person per decennial survey; the advantage would be that occupational mortality rates could be estimated without bias, which is not at present possible (since at present the occupation reported by the informant to the registrar of deaths is compared with the occupation described in the decennial census, and these are not compatible).

Proc. R. Soc. Lond. B. **205**, 179–197 (1979)

Printed in Great Britain

Hazards from chemicals: scientific questions and conflicts of interest

By A. E. M. McLean

*Laboratory of Toxicology and Pharmacokinetics, Department of
Clinical Pharmacology, University College Hospital Medical School,
University Street, London WC1E 6JJ, U.K.*

All substances are toxic when the dose is large enough. In order to
regulate the use of chemicals, we need to measure the level at which
toxic effects are found. Epidemiological evidence suggests that present
levels of chemical use do not lead to widespread harmful contamination
of the human environment. For chemicals, most of the problems of
toxicity are found in the workplace, while the population at large gets
most of its toxic effects from voluntary exposure to substances such as
tobacco smoke and ethanol. The prevention and control of toxic effects
depends on a series of steps. This begins with measurement of toxicity
in model systems, such as laboratory animals, and the estimation of the
likely exposure of workers or consumers. Reliable extrapolation of
information gathered from animals to the diverse and biochemically
differing human population depends on understanding mechanisms of
toxic effects.

The toxic effects and mechanisms of action of substances such as
carbon tetrachloride or paracetamol have been extensively investigated, and our ability to predict toxicity or develop antidotes to poisoning
has had some success, but epidemiology is still an essential part of
assessment of toxic effects of new chemicals. The example of phenobarbitone shows how animal experiments may well lead to conclusions
which do not apply to man. After measurement of toxicity and assessment of likely hazards in use comes the final evaluation of the use of
a chemical. This depends not only on its toxicity, but also on its usefulness. The direct effects on health may be small in comparison with the
indirect advantageous effects which a useful substance such as vinyl
chloride may bring.

The assessment of risks and benefits of new chemicals can be partly
removed from a political style of discourse, but the evaluation of the
relative weight to be attached to these risks and benefits is inescapably
political. The scientific contribution must be to allow the debate to take
place in the light of maximum clarity of information about the consequences of use of chemicals.

Introduction

Modern societies depend on the extraction, manufacture and use of materials
ranging from natural products such as asbestos or coal to synthetic organic
compounds like vinyl chloride, unknown in nature. The scale of use and transport
varies from millions of tonnes of petroleum hydrocarbons and chlorine, to a few

tonnes of insecticides and kilograms of drugs and food colours (Korte 1977). These diverse substances form a new chemical environment for man. The possibility that they may have adverse effects on the human populations and the natural environment, not previously exposed to the materials, has gradually become part of public consciousness (Carson 1963; Kates 1978; Jones *et al.* 1978; Royal Commission on Environmental Pollution 1976). The question is: how can we evolve a system that deals with the new dangers of this new environment, while retaining the more beneficial aspects of technology?

It is best to separate out these political and scientific questions into their component parts (Lundqvist 1977; Gregory 1971). Is there evidence of harm from industrial chemicals now? How can we examine new chemicals to predict what effects their future use would bring? What institutions can we set up to decine whether and how to use new chemicals?

SOCIAL AND CHEMICAL FACTORS IN THE CAUSE OF DISEASE

There is massive evidence that many chemicals used in industry have serious long-term toxic effects on some workers in industry, and that strenuous and continued efforts are required to reduce exposure of workers to toxic chemicals (see Hunter (1975) or any issue of the *British Journal of Industrial Medicine*).

The pulmonary diseases of asbestos workers and the increased risks of mental impairment or death from coronary heart disease among rayon workers exposed to CS_2 are two examples among very many (Elmes & Simpson 1977; Hänninen 1971; Tiller *et al.* 1968). It is difficult to assess the scale of injury, because industrial workers are not a random cross section of the population, but are markedly healthier than average at the time of entry to work, so that survival, illness and disability will depend on the kind of person who enters the industry as well as the conditions of work (Fox & Collier 1977). However, mortality among most occupational groups seems more related to social class than to the particular materials that are handled at work. For instance, unskilled workers in the various industries from catering to chemicals do not have greatly differing mortalities, though they are all high in comparison with skilled or professional workers (Registrar General 1971, 1978). This may reflect generally dusty, cold or wet conditions of work and life, or reflect a way of organizing living conditions that is not conducive to longevity (Knowles 1977; Belloc 1973; Brown & Harris 1978; Brown *et al.* 1975). When the survival of the wives and children as well as men at work is considered, it becomes clear that there is a steep social gradient of mortality and presumably morbidity, most of which cannot be due to exposure to the chemicals found in the workplace (tables 1 and 2). Table 1 shows that almost a third of warehousemen die in the 20 years before reaching retirement age. In this and in the causes of death they are typical of the general population (standardized mortality ratio (s.m.r.) 104 instead of 100). Of this group 12 % will die of lung cancers, one might say, voluntarily. They differ markedly from teachers who have almost halved their mortality in the age range 45–64 years. Both the leukaemia s.m.rs are close to the expected 100 value, showing that there are unlikely to be gross errors of estimate of population size.

TABLE 1. MORTALITY AMONG TEACHERS, WAREHOUSEMEN AND THEIR WIVES

(Deaths per annum per 100000 persons.)

age	teachers (wives)	warehousemen (wives)
25–34	73 (48)	115 (57)
35–44	135 (106)	248 (155)
45–54	414 (277)	771 (381)
55–64	1297 (617)	2223 (973)

standardized mortality ratios

	teachers (wives)	warehousemen (wives)
all deaths	60 (66)	104 (98)
leukaemia	102 (64)	93 (90)
lung cancer	34 (64)	107 (101)
coronary heart disease	80 (60)	110 (96)

From Registrar General (1971), tables 3Aii and 3Bii; International Classification of Occupation nos. 287 and 210.

TABLE 2. INFANT MORTALITY 1972

(Deaths per annum per 1000 live births.)

London		Hampshire (rural)	14
Tower Hamlets	26	Hampshire (Portsmouth)	15
Lambeth	19	Anglesey	22
Camden	15	Yorkshire (towns)	21
Sutton	13	Yorkshire (rural)	19

From Registrar General (1974), table 13.

The wives of teachers and of warehousemen have lower mortality rates than their husbands, but again show a strong social gradient. Table 2 shows that infant mortality is high in the poor areas of London (Tower Hamlets and Lambeth) while Camden, being more prosperous and with an unusual concentration of major hospitals and paediatric services, does better in spite of being in the central area of the town. The other figures also suggest that prosperity and good health services are the important shields against death in infancy.

The Study Group on Long-term Toxic Effects to Man from Man-made Chemicals was set up in 1975 to consider the question whether there was any widespread danger to people from chemicals in the environment. The answers seem to be that there are localized patches of harmful pollution, such as areas with excessive lead in the drinking water in parts of Scotland, excessive dust and sulphur dioxide in the atmosphere in some towns, and asbestos dust outside factories, but that there was no evidence of present widespread danger to health from environmental pollution, outside the workplace.

We look at public health records and we see that almost 30% of men die between the ages of 45 and 64 years, and that the incidence of a number of cancers and other serious diseases is not falling. There is powerful evidence that the major differences in disease and mortality between nations and between social classes are environmentally and socially determined and are not genetic in origin. The question is then forced on us: what aspect of our environment causes these differences in disease incidence?

The major causes of premature death are accidents, coronary heart disease, and the cancers. The present role of specific man-made chemicals in the causation of these ills seems to be a minor one, since the variation between town and country, and between occupational groups of similar social standing but handling different materials, are, with minor exceptions, small. The major differences between groups seem to be largely accounted for on the basis of differences in cigarette consumption, exercise, diet and all of the socially determined living conditions, from overcrowding to the availability of bathrooms (Registrar General 1978; Brown & Harris 1978).

We still need to investigate specific chemicals, by using the techniques of clinical observation, epidemiology and toxicology to measure effects of chemicals in use, and in particular we need to prevent future errors such as the introduction of cigarettes some 70 years ago.

In addition, the occupational and accidental hazards from exposure of relatively small groups of people to reactive chemicals are likely to increase. Such groups of a few hundred people per group will be too small for most epidemiological work, and we will depend on the accuracy of our predictions of toxicity before human exposure takes place.

One of the first steps is to try to predict whether the 500 or so new chemical compounds that are brought into commercial use each year are likely to have harmful effects, and to investigate those existing chemicals about which there is suspicion.

PRESENT METHODS OF CONTROL OF NEW CHEMICALS AND PREDICTION OF ADVERSE EFFECTS IN USE

When a manufacturing company develops a new compound, say an insecticide to control carrot fly, this is done on the basis that there is a need for the product and a possible market that will justify the expenditure of resources in development (but see Galbraith (1975) on creation of needs). Before the material can be applied to a crop, even on an experimental basis, clearance from the controlling committee is requested (in the U.K. the scientific subcommittee of the Pesticide Safety Precautions Scheme at the Ministry of Agriculture, Fisheries and Food would be approached; M.A.F.F. 1971).

All substances are toxic, given a sufficient dose. To ask 'is this substance toxic?' is not the right question; oxygen, water or vitamin D are toxic. Toxicity is a property of materials, like density. The question that we need to ask is whether there is risk attached to the use of a material, and this will depend on the quantitative and qualitative aspects of the toxicity of the material and its particular use.

We can measure toxicity in model systems, and by epidemiological techniques. When our hypothetical insecticide is presented to regulatory committees it has already a file of 'toxicity data' of perhaps 1000 pages.

TABLE 3. A GENERAL STRATEGY FOR ASSESSMENT OF RISKS FROM NEW CHEMICALS

1. Toxicity: measured in model systems.
2. Risk assessment: from toxicity measurements and exposure estimates.
3. Risk: evaluation from assessments of risk and benefit and the alternatives available.
4. Epidemiological: surveillance to guard against unsuspected risks and as verification of evaluation procedures.
5. Responsibility and resources for safe handling assigned.
6. Reiteration of steps 1–5 in light of experience. Each step is considered for (*a*) worker, (*b*) environment, (*c*) consumer.

TABLE 4. HEALTH AND SAFETY COMMISSION PROPOSALS

(Part of proposed minimal list of information to be requested by Health and Safety Executive for registration of all new chemicals.)

1. Chemical and physical properties
2. Biological properties:
   acute oral or inhalation toxicity;
   eye and skin irritancy;
   results of short tests for carcinogenicity, mutagenicity and teratogenicity;
   subacute toxicity (30 days);
   biodegradability;
   fish toxicity.

(From Health and Safety Commission 1977.)

The effects of single and multiple doses, given by various routes (to various species of animals), will be described. The teratogenic, mutagenic and carcinogenic potential of the chemicals will have been investigated, together with effects on skin and eye, on birds, bees and fish, on reproduction and on bacteria. The metabolism, excretion, and perhaps mode of action of the compound will be presented. All this is usually done at dose levels far higher than the doses to which operatives or consumers will be exposed. The doses are designed to demonstrate what kind of toxicity the compound displays, which tissue is affected, and what functions altered. The mass of information is presented, together with much other material, to a committee of medical, agricultural and wildlife specialists who may ask for more investigations to be done (M.A.F.F. 1971).

From this array of experimental results on model systems emerges a general picture of the ability of the compound to cause harm to living organisms. In particularly, the dose–response relation for different effects is observed. The compound can then be compared with other compounds of similar structure, or similar action, whose effects on man and environment are known from previous use, or often from previous misuse and regretted ill-effects.

Given our assessment of the toxicity of the new compound (tables 3 and 4), we then look at the proposed use, and the estimated exposure of different groups of people such as agricultural workers inhaling the spray, or consumers eating residues on food. From the dose predicted for man and from the effects on model systems we make predictions of what is likely to happen in use. At this stage it is possible to make an evaluation, a value judgement, of whether the benefits make the risks acceptable.

In essence and to a variable extent, some such procedure is followed for new drugs, pesticides and food additives, and will soon have to be followed for all new industrial chemicals (Health and Safety Commission 1977; D.H.S.S. 1977) The ability to predict whether a new substance will cause harm in use is imperfect. First, there are massive variations in the way that different species and different individuals respond even to the same measured dose of a compound. The calculated dose to which man is exposed rarely takes into account inter-individual variations of dose due to variation in behaviour. A lady who eats 1 kg of carrots per day recently appeared in hospital because of the yellow colour of her skin, from carotene, not from pesticide residues. Should we allow for such individuals? No one would have predicted that some men would smoke 80 cigarettes per day when these were first marketed, nor that some individuals would take kilograms of phenacetin headache powders. We need information on the extent to which instructions to wear protective clothing when spraying insecticides are complied with.

Another major difficulty comes from accidental, improper, or even illegal use of new substances. Insecticides spilt in railway wagons have contaminated sacks of flour. Jet engine oil has been used to adulterate cooking oil, and dangerous materials are frequently used improperly in the factory or farm (McLean 1972; Jones et al. 1978; Mark & Stuart 1977; Rycroft et al. 1976; Smith & Spalding 1959).

Lastly, we have the problem of prediction of the effect of a substance such as a food colour on large numbers of individuals, perhaps millions, when we have a dose–response curve from animal models or human studies, where at the most only a few hundred individuals were observed.

This last question is particularly disquieting where we have an 'all or none' type of effect and the response is measured in terms of probability of effect versus dose of substance. This is so for induction of cancer by chemicals and leads to a difficult calculation of low probability multiplied by large numbers of persons exposed.

Because of the difficulties of prediction of 'safety in use', the large number of new drugs and industrial and agricultural chemicals being developed, and shortage of toxicologists, many regulatory authorities and industrial scientists have tried to set down a detailed scheme of tests for the assessment of toxicity of new chemicals. If a set of tests could be prescribed, then the developer firm need only carry out the work on the prescribed list, and the regulatory authority can tick off the results on a check list. The process of 'toxicity testing' would become a simple predictable routine, with a predictable cost and time course for any new compound. Unfortunately, it seems impossible to devise a set scheme which will give us the information we need. We can only give general principles of investigation, and make a balance between laboratory studies before use and surveillance of exposed populations after use has begun (M.A.F.F. 1971; D.H.S.S. 1977).

Since an infinite number of investigations are possible, it is necessary to decide which of these are relevant in the light of the nature and intended use of the chemical. It is like the investigation of a new disease. No routine set of steps, no rigid guidelines of investigation can be laid down for the investigator, except for the very minimum that needs to be done. For instance, the acute toxicity to mammals in terms of lethal dose, and organ affected, should be known. Some estimate of persistence in biological systems, mutagenicity, and effects of repeated exposure will be needed. But depending on the results of the first investigations, the next steps must be designed to lead to elucidation of mechanisms. The alternative way of 'routine' investigation already leads to massive, uninformative, wasteful, and misleading 'toxicity tests'. The quality of the investigation falls and scientists of imagination, needed to solve new problems of new chemicals, are discouraged from this field.

MECHANISMS OF TOXICITY AND EXTRAPOLATION TO MAN

If we know the way in which a new toxic material penetrates metabolic paths and the molecular targets with which the toxic reacts, to cause injury, we are then able to use our knowledge of the biochemistry and physiology of any species to predict responses to the new toxic material. In contrast, the crude and mechanical description of, say, lethal effects in one species will tell us little about other species, or about variation inside a species. Human populations have great genetic diversity, contain individuals of different ages and states of health, living in a wide variety of environments. In particular human diets vary greatly, especially in regard to the amount and type of foods eaten. How can we take these factors into account?

EXAMPLES: PARACETAMOL AND CARBON TETRACHLORIDE

The drug paracetamol (*N*-acetyl-*p*-aminophenol) is a useful analgesic, but causes liver necrosis and death in rats or man when taken in overdoses of from 20 to 100 times the normal dose. The question can be asked; is this an adequate safety margin, or will some individuals be sensitive to normal or repeated doses?

The drug is well absorbed, and is normally removed harmlessly after conjugation in the liver to a glucuronide or sulphate. Toxicity depends on a second pathway of metabolism by drug oxidizing enzymes centred on cytochrome P450 in the endoplasmic reticulum of the liver cell. A reactive metabolite, probably N-hydroxyparacetamol, is formed, and combines with glutathione. As the liver reserves of glutathione fall, so the metabolite attacks other cell components and cell death results. Treatment with inducers of synthesis of the cytochrome P450 enzymes, such as DDT or phenobarbital, or treatment with low protein diets that reduce liver glutathione content, make animals sensitive to paracetamol (Mitchell *et al.* 1975; McLean & Day 1975).

TABLE 5. THE EFFECT OF DIET AND PHENOBARBITAL TREATMENT ON LIVER CELL COMPOSITION ON THE LETHAL EFFECTS OF PARACETAMOL AND CARBON TETRACHLORIDE IN RATS

diet	pheno-barbital treatment	liver glutathione µmol/g liver	liver cytochrome P450 nmol/g liver	l.d.$_{50}$, paracetamol g/kg	l.d.$_{50}$, CCl$_4$ ml/kg
stock	−	6.9 ± 1.7	40 ± 9	5.2	3.6
low protein	−	2.2 ± 0.2	23 ± 5	2.1	14.7
stock	+	7.6 ± 0.1	142 ± 35	2.0	0.5
low protein	+	2.8 ± 0.3	81 ± 15	0.9	1.4

(Data from McLean & McLean (1966), Garner & McLean (1969), McLean & Day (1975) and unpublished results.)

Knowing the normal range of variation of cytochrome P450 and glutathione content of human liver, and from epidemiological studies of overdose, and making reasonable assumptions about rates of synthesis of glutathione and the range of dietary protein intake, we can say that the usual 1 g dose of paracetamol will not cause liver damage to any member of the population even if repeated at 4–6 h intervals, and even if used in a developing country. However, we can never rule out the possibility that a genetic variant may exist where differences in metabolism or immune systems cause individual sensitivity to this or any other chemical.

In contrast to paracetamol, laboratory investigations on the relation of diet to toxicity of carbon tetrachloride (CCl$_4$) showed that toxicity and glutathione levels were not related, but the amount of tissue injury and lethal effects produced by a dose of CCl$_4$ were directly proportional to P450 levels in the liver (table 5). We can conclude that the toxic metabolite of CCl$_4$ is not detoxicated by glutathione and it is worth considering whether the metabolite exerts its effects entirely in the lipid phase of the cell.

By altering the chemical environment of an animal we alter its cell composition in an adaptive manner; this in turn can alter response to toxic materials, by an order of magnitude. The conclusion that one may draw from experiments of this kind is that the assessment of the toxicity of a compound, based on the effects of exposure of healthy laboratory animals fed stock diets, is a very limited assessment. If we can improve our understanding of mechanism, then our extrapolations become more confident; the use of methionine in therapy of paracetamol poisoning was based on such experiments (McLean 1974). Human epidemiology remains a necessity for rational control of potentially toxic chemicals, because we cannot extrapolate with certainty from animal models.

EPIDEMIOLOGY TO VERIFY TOXICITY ASSESSMENT

In man, acute exposure to CCl_4 leads more often to renal injury than liver damage. The kidney also contains enzymes linked to cytochrome P450, but the reason for the distribution of injury between liver and kidney in rat and man is unknown.

Even when we do have a reasonable idea of a mechanism of toxicity, we still have to monitor the effects of human exposure to new chemicals.

Without such studies on man, we can never verify the correctness of our techniques of measuring toxicity in model systems. We may be using quite inappropriate models and unless we correct our techniques we could go into an irrational system of regulations. Chemicals that cause mutations in bacteria or *Drosophila* are used in many industrial processes. There are no known instances of increased mutations in the offspring of persons exposed to mutagenic chemicals. Even though this is a negative finding whose reliability depends entirely on how many people are examined, such knowledge is the only way of providing a sense of perspective for judging the relative importance of regulations against mutagenic chemicals (except in so far as they may also pose carcinogenic hazards) versus regulations designed to reduce acute poisoning of persons using insecticides, or, say, road accidents. In dealing with toxic chemicals, errors that underestimate risks will injure workers, consumers or environment. Errors that overestimate risks will also cause damage because the unnecessary abandonment of, say, a new drug will lead to loss of the expected benefits, and may lead to the loss of the research and development costs. The loss of great investments of time and materials will eventually be spread over the society, and make resources unavailable for other objectives (Westlake 1977).

The required epidemiological work can take two forms. First, there can be surveillance of the physical and mental health of the exposed and control populations with the use of questionnaires and general practice records. Secondly, there can be specific surveys relating to items of suspicion, perhaps to fears that some specific tissue will be injured, arising from toxicity data. For instance, liver function tests might be carried out in an exposed population.

PHENOBARBITAL AND THE WRONG MODEL SYSTEM

In assessing the potential of a compound to cause cancer in man, we rely largely on animal experiments where large doses of the compound are fed to rats or mice.

The class of substances that induce synthesis of cytochrome P450 and the associated drug metabolizing enzymes in the liver of most mammals includes phenobarbital, DDT, and many other drugs and chemicals, some of which occur naturally. Some of these are powerful inducers, causing an increase in liver mass and cell number as well as an increase in size of the hepatocytes. Long-term feeding with phenobarbital or DDT leads to an increased incidence of liver tumours in some strains of mice and rats, and some of these tumours are malignant. There is no doubt that for certain strains of mice phenobarbital is a carcinogen. The question is whether the classification of compounds into 'carcinogens' and 'non-carcinogens' is a rational classification, or whether it is a misleading over-simplification. The development of a tumour depends on multiple factors in the host and the environment (Anon. 1974) and several of these factors are necessary but not sufficient causes. For instance, high protein and high fat diets cause a marked increase in incidence of both spontaneous and carcinogen-induced tumours. If all inducers of microsomal enzymes are to be called carcinogens, we have to face the fact that many natural compounds and many pesticides and drugs are inducers, and that increasing the dietary protein and fat content has similar effects.

The discussion is made more intense and more difficult by the observation that phenobarbital treatment can protect against the carcinogenic effects of aflatoxin on rat liver (McLean & Marshall 1971) and that two economically important insecticides, aldrin and dieldrin, seem to produce liver tumours by the 'induction' type of mechanism.

It becomes important to decide whether the human liver is like the mouse liver in its response to inducing substances, and since no mechanism for the carcinogenic effect in the mouse has been discovered, we have no theoretical basis in which to work.

However, we are fortunate to have a large human population to study, since phenobarbital has been used as an anti-epileptic drug for over 40 years (Clemmesen et al. 1974).

Studies in Denmark and in London have shown no excess of liver tumours among patients with epilepsy (table 6). The incidence of liver cancer in the general population of England and Wales is very small (28 per million in men aged 55–64 years, in comparison with 2900 per million for lung cancers, or an estimated rate of 1070 for liver cancer in an area of Mozambique in which food is heavily contaminated with aflatoxin) (Van Rensburg et al. 1974), and in any sample there is the possibility of missing the patients who have the tumours. In our sample there is a 7 % probability that we would miss even an eightfold increase of liver cancer incidence. However, such an increase would still make the risk from liver cancer negligibly small in comparison with the risks of death from untreated epilepsy, with the risks from cigarette smoking, or with the benefits from the use of phenobarbital.

Since phenobarbital acts as an effective inducer of P450 synthesis only on liver tissue, we can say that any carcinogenic risk to man from this whole class of substance is unlikely to affect other tissues and is very small in comparison with the normal risks of living. The model system of tumours in the mouse liver has probably given a false positive signal, while the studies on patients with epilepsy lead us to look at other factors that cause the high observed mortality.

TABLE 6. MORTALITY IN 2000 PATIENTS WITH SEVERE EPILEPSY
(CHALFONT CENTRE), 1951–77

	observed	expected	O/E	95% confidence limits for O/E
deaths, all causes	636	208	3.1	2.8–3.3
liver cancer	1	0.6	1.5	0.0–8.5
accidents	64	14.6	4.4	3.3–5.6
epilepsy (mostly status epilepticus)	207	2.7	75.8	65.7–87.0
cancer, all sites	73	51.6	1.5	1.2–1.9
lung cancer	23	16.3	1.4	0.9–2.1

(Study by S. White & A. McLean, manuscript in preparation.)

There are many other technical problems in the assessment of toxicity of chemicals. There is difficulty in detecting the potential for renal toxicity and bone marrow toxicity. The ability of a compound to produce skin sensitization and other allergic diseases in man is hard to assess from investigations in animals.

Each new investigative technique poses difficulties of interpretation. For instance, practolol metabolites seem to bind to proteins and so act as haptenes in producing an immune reaction. The relation of such an immunological effect to the disastrous side effects of practolol is unknown, so we cannot use the technique to investigate new compounds with any confidence (Amos *et al.* 1978).

Each new technique has to be applied to a large series of compounds which have already been used, so that we can see whether, say, metabolite binding to plasma proteins correlates with delayed side effects, or whether it is an incidental event of no consequence and no utility in the investigation of new compounds.

These are questions of the design and interpretation of experiments to investigate the toxic effects of new compounds. There are other questions of equal importance concerning the design of institutions and procedures to control the use of toxic chemicals.

CONFLICTS IN THE PERCEPTION AND CONTROL OF RISKS

We cannot eliminate or even detect all of the risks to life and health that come to workers in their workplace, or to consumers in their use of products. This is first because every single substance carries risk in its manufacture and use, from drowning in beer and cancer from asbestos to falling from household ladders. Secondly, we can detect only those events that are sufficiently noticeable or frequent to show up in studies on exposed and control populations. Sensitivity to a chemical due to a rare metabolic pathway would not show up in human or in animal studies unless it caused a bizarre illness.

Our present model systems for assessing risks or benefits do not give us sufficiently precise answers for us to be able to predict the effects on quality of life of the introduction of a new product, or consequences of a ban on use of an old chemical. A new drug might be a minor advance or a major disaster. Stringent controls on a chemical such as benzene may save lives at a huge cost to the community, or may save no lives but lead to improved production and management methods, or any combination of effects (Westlake 1977).

Since all processes and products carry risk it is no great advance to say that we must permit only acceptable risks. The questions are: what is an acceptable risk, and who decides on its acceptance?

No expert is ever at a loss for more tests that he thinks could be done to define the toxicity of a chemical. The protection of the environment and prevention of risks to health of workers and consumers are put forward as an absolute, good, and sufficient reason for more testing and more precautions. Each expert presses the urgent need for work in his own field, be it neurotoxicity, cancer, mutagenesis or damage to bees.

In committee one sometimes gets the feeling that no one without a degree in toxicology should be allowed to take a bath lest there be a side effect damaging health or environment.

Meanwhile there are groups in industry who seem incredulous at the idea that any product they make could ever do any significant harm, and see no need for government committees to do anything except endorse the safety of their products.

Outside these official circles are critics who say that testing is inadequate, companies venal, committees ignorant and corrupted (Kinnersley 1974; Gillespie et al. 1978; McLean 1978).

We could spend a large proportion of the national product on protection of the environment and safety testing procedures, but might rapidly reach the point where the loss of production and lowered living standards depressed health far more than any gain coming from increased knowledge about toxicity.

There are several conflicting interests involved in the process of deciding how much of the society's resources should be devoted to ensuring safety and who should pay. The conflicting interests of owners of enterprises and the workers who carry risks are obvious, but need to be resolved. Making available the best scientific information to both parties may help. Certainly secrecy has no proper place in the field of toxicology, for only by open disclosure of toxicity data and risk estimates can proper evaluation take place (Smith 1977).

Government action to ensure that all enterprises have to comply with the same proper safety standards will help to ensure that managers are less tempted to secure competitive advantage by skimping safety costs.

The conflicts over distribution of advantages gained by new economic processes between worker, management, owners and the rest of society also spill over into discussion on safety. The interests of the rest of society can appear to be contrary to those of the workers in a particular industry. For instance, there are over 150 accidental deaths per year on building sites in the U.K. (H.M. Chief Inspector of Factories 1974). The better safety standards that ought to be imposed would raise building costs, as well as requiring a bigger inspectorate to oversee the sites. One could argue that the major source of improvement in public health has been improved living standards, in the form of better housing, food, and social services. These benefits are possible because of increased productivity of labour, and if increased efforts in the direction of safety are misplaced, then the health advantages made possible by resources created by a new technology would be lost.

At times it may be right and necessary for social reasons to incur safety costs that are not cost effective in comparison with investment in some other health field. For instance, the cost of bringing vinyl chloride monomer concentrations down from about 20 parts/10^6 to 4 parts/10^6 or less, at a cost of about £10000 per exposed man, will probably be far less effective in saving life than a similar sum spent on the more dangerous building sites (Lassiter 1977). However, the known carcinogenic risk, although much less than that posed by cigarettes, is regarded as intolerable perhaps because it is insidious and incurred at work, and so regarded as a risk imposed by management rather than undertaken voluntarily. The economic impact of improved safety regulations and their enforcement, onto general living standards should be studied. It is possible that increased attention to safety would improve managerial involvement and improve industrial relations if correctly planned. Just as we have a budget for the national health services inside which we try to allocate resources, so we need to decide on a budget for environmental quality and risk avoidance. Inside this broad field we could allocate resources to specific areas, like noise in industry, or risk assessment of new drugs. Unlike health service expenditure, which is centrally funded, much of the allocation of resources on safety comes in the form of requests by government committees that someone else, usually industry, should spend money, find and train people, or feed and examine a few thousand rats. Such requests do not appear on the public expenditure bill, but are still allocations of resource and perhaps it is time that we acknowledged that these resources are limited.

This proposal, of making a budget for risk avoidance, poses considerable difficulties, but so does the present alternative, where only informal commonsense stands in the way of an infinite expansion of requirements for testing of chemicals, drugs and pesticides. We could start by taking one field, say the evaluation of food additives, and making a budget to allocate expenditure between animal tests, and observations on human populations. It is doubtful whether the present division of resources in research on, say, nitrosamines in food would still be heavily biased towards rat experiments if such a budgeting procedure were followed.

ACCEPTABLE RISK

The concept of acceptable risk is currently much discussed (Lowrance 1976; Starr 1969; Starr et al. 1976; Kletz 1976, 1977; Ashby 1976; Pochin 1975; Farmer 1975, 1977; Fairclough 1977; Council for Science and Society 1976, 1977). Acceptable risk is not definable as a particular quantity. People undertake voluntarily the high risks of cigarette smoking, rock climbing, or driving motor cars. At the same time, involuntary risks imposed by lead in petrol or food additives are accepted only reluctantly and only if the risk can be described as unmeasurably small. The risks at work seem to be more intolerable if due to a chemical with carcinogenic or other insidious toxic effects, than if the risk is posed by a visible feature such as the danger of falling from scaffolding. Similarly, surgical risks are accepted more readily than medical risks (Glaser 1977). Certainly the pressure for regulation of chemical hazards seems far greater than the pressure against causes of mechanical trauma, which may be just as much a cause of permanent disability (Kinnersley 1974). The distinction may lie not in the magnitude of the hazards but in the objections which people have to being controlled and used by other persons. So, lead in the air is regarded as an imposition on the public by the actions of motorists and oil firms, while nitrate in drinking water, present in amounts closer to toxic levels, is regarded with equanimity perhaps as a natural consequence of agriculture. Drug toxicity is often regarded as intolerable. Acceptability of risks is essentially a political problem of accommodation between different groups in the society, with separate interests, depending on occupation, age, education and many other factors.

The perception of risks is frequently inaccurate (Abelson 1977; Slovic et al. 1976) and again depends on social factors, so farmers and scientists working in pesticide manufacture see the balance of risk and benefit from pesticide use quite differently from the way in which independent pesticide specialists and wildlife specialists see those same benefits and risks (Davis 1978).

An accident in which six people are killed at once is regarded as 'newsworthy' and more serious than six separate fatal accidents, or the background figure of 15 motor traffic deaths per day in the U.K. Perhaps this reflects the more serious social consequences of multiple deaths in one family or community, in comparison with separate events spread over many social groups. Similarly, 300 men each off work for 1 day because of minor illnesses are quite different from one man losing 300 work days, so adverse effects cannot be reduced to a simple currency of days off work.

Joshua Lederberg (1974) pointed to the social loss implied in setting the risk levels so low that new compounds are not developed and that the effort expended in examining new compounds should be rationally examined and set to produce a benefit: cost ratio that is socially acceptable.

The University of Sussex group have suggested that safety costs were unevenly distributed between industries, but make the assumption that such costs are direct costs of production without any benefit to management or to productivity. When the costs of air-conditioning are treated as 'safety costs' the argument becomes dubious (Sinclair *et al.* 1972).

The Science Council of Canada (1977) has taken the discussion further than other groups, and sees the key to the problem as being informed discussion between the representatives of the major groups, consumer, worker, and management, leading to agreement on action to be taken. It is becoming clear that no experts can decide what is an 'acceptable' risk. Expert groups can only decide what the risks are.

The Science Council recommends the setting up of a unified 'Advisory Council on Occupational and Environmental Health'. This body would have the responsibility to commission research to define hazards, and then to set in motion procedures to control the hazards.

In the U.K., the plans put forward by the Health and Safety Executive propose that when considering new chemicals in their industrial aspects, trade union representatives should be consulted. But frequently the consumer's voice is also needed when arrangements are being made between the large organized bodies, departments of government, manufacturing firms and trade unions. For instance, many consumer products such as wood preservatives, over-the-counter medicines, or the caustic soda used to clear drains, are hazardous if used carelessly, as are paraffin oil, matches, or axes. The temptation is for official bodies to restrict the use of any substance to which a hazard can be attached, to some professional licensed group. So some timber preservatives are on sale only to specialist firms. Since everyday life has hazards, we should recognize some chemicals as posing risks of only average severity which do not increase the overall risks to life by an appreciable amount and permit their use in the household, provided that adequate labels are supplied.

It has become clear that our increasing ability to analyse information about mortality and disease gives us new power to detect risks. New analytical methods enable us to detect environmental contamination, new toxicological methods allow us to infer that hazards of carcinogenesis, teratogenesis and other tissue injury are associated with particular chemicals.

We are unable to guarantee safety of new drugs or other chemicals, but the demand for safety grows together with the demand for new products. At the same time, 'voluntary' risks, such as those posed by cigarette smoking, driving motor vehicles, or inactivity are responsible for a high proportion, perhaps half, of all deaths before retirement. We are unwilling to discontinue the use of all hazardous chemicals, because of the drastic decrease in consumption that this would involve. Since we accept the use of toxic chemicals knowing that they present a risk, what conditions should we make before permitting them? Table 7 puts forward a suggested list.

TABLE 7

A risk to life and health is unacceptable unless it is minimized, or if unavoidable, equitably spread through society.

Except for risks that are:
  (1) voluntary, not imposed;
  (2) known, not concealed;
  (3) benefit is shared by the risk taker and society;
  (4) information on consequences of risk is measured and made public to reduce further risk.

In the cases of deep sea fishing, immunization of children, or vinyl chloride polymerization, the conditions for 'voluntary risk' are now more or less fulfilled. With cigarette smoking the question of benefit and cost is more doubtful, since the cost to society may be much larger than current benefits (Atkinson & Townsend 1977). For food additives and colours and pesticide residues in food, the element of concealment and lack of any epidemiological surveillance makes the present style of use doubtful.

LEGAL RESPONSIBILITY

A further step is to assign responsibility for management of toxic chemicals.

Societies regard certain commodities as precious or dangerous and set aside particular individuals and procedures to guard these substances. Explosives, morphine, radioactive substances, gold or money fall into this category (Council for Science and Society 1977).

In each case there is a dosage factor: small amounts are permitted for personal use, from coins to percussion caps or luminous watch dials. Larger amounts are in the care of responsible individuals, such as accountants or hospital physicists, who have to account for the proper handling and disposal of money and radio-isotopes respectively. Perhaps we need toxic substance managers who are qualified to accept delivery of designated toxic substances, and are held responsible for safe disposal. Almost any substance can be dangerous, as even flour dust can explode. The problem is to combine knowledge of toxicity, perception of hazard, and motivation to carry out proper management. The scientists' task is to obtain and organize the information about risk so that it becomes accessible and under-standable to the people at risk and to those responsible for minimizing risk (Siekevitz 1970).

References (McLean)

Abelson, P. H. 1977 Public opinion and energy use. *Science, N.Y.* **197**, 1325.

Amos, H. E., Lake, B. G. & Artis, J. 1978 Possible role of antibody specific for a practolol metabolite in the pathogenesis of oculomucocutaneous syndrome. *Br. med. J.* **1**, 402–404.

Anon. 1974 Does this chemical cause cancer in man? *Lancet* ii, 629–630.

Ashby, E. 1976 Protection of the environment. The human dimension. *Proc. R. Soc. Med.* **69**, 721–730.

Atkinson, A. B. & Townsend, J. L. 1977 Economic aspects of reduced smoking. *Lancet* ii, 492–494.

Belloc, N. 1973 Relationship of health practices to mortality. *Prev. Med.* **2**, 67–81.

Brown, G. & Harris, T. 1978 *Social origins of depression*. London: Tavistock.

Brown, G. W., Bhrolchain, M. & Harris, T. 1975 Social class and psychiatric disturbance among women in an urban environment. *Sociology* **9**, 225–254.

Carson, R. 1963 *Silent spring*. London: Hamish Hamilton.

Clemmesen, J., Fuglsang-Frederiksen, V. & Plum, C. M. 1974 Are anticonvulsants oncogenic? *Lancet* ii, 705–707.

Council for Science and Society 1976 *Superstar technologies*. London: Barry Rose Publishers.

Council for Science and Society 1977 *The acceptability of risks*. London: Barry Rose Publishers.

Davis, J. E. 1978 Cost benefits of transfer of cancer legislation on the industrial and agricultural segment of developing countries. *Proc. Cancer Workshop*, March 1978. Lyons: I.A.R.C. (In the press.)

D.H.S.S. 1977 *Notes on applications for Clinical Trial Certificates*. London: Department of Health and Social Security.

Elmes, P. C. & Simpson, M. J. C. 1977 Insulation workers in Belfast. A further study of mortality due to asbestos exposure (1940–1975). *Br. J. ind. Med.* **34**, 174–180.

Fairclough, A. J. 1977 Some aspects of environmental risk assessment as seen from the standpoint of policy workers and decision takers in central government. In *Proceedings of Seminar on Environmental Risk Assessment*. Tihanyi, Hungary, June 1977. (SCOPE Project 7).

Farmer, F. R. 1975 Accident probability data. *J. Inst. nucl. Engrs* **16**, 1–13.

Farmer, R. 1977 Today's risks: thinking the unthinkable. *Nature, Lond.* **267**, 92–93.

Fox, A. J. & Collier, P. E. 1977 Mortality experience of workers exposed to vinyl chloride monomer in the manufacture of polyvinyl chloride in Great Britain. *Br. J. ind. Med.* **34**, 1–10.

Galbraith, J. K. 1975 *Economics and the public purpose*. London: Penguin Books.

Garner, R. C. & McLean, A. E. M. 1969 Increased susceptibility and carbon tetrachloride poisoning in the rat after pretreatment with oral phenobarbitone. *Biochem. Pharmac.* **18**, 645–650.

Gillespie, B., Eva, D. & Johnston, R. 1978 A tale of two pesticides: why is diledrin carcinogenic in the U.S. but not in Britain? *New Scient.* **77**, 350–352.

Glaser, E. M. 1977 Acceptable and unacceptable risks. *Br. med. J.* ii, 1028–1029.

Gregory, R. 1971 *The price of amenity: five studies in conservation and government*. London: Macmillan.

H.M. Chief Inspector of Factories 1974 *Annual Report*. London: H.M.S.O.

Hänninen, H. 1971 Psychological picture of manifest and latent carbon disulphide poisoning. *Br. J. ind. Med.* **28**, 374–381.

Health and Safety Commission 1977 *Discussion document. Proposed scheme for the notification of the toxic properties of substances*. London: H.M.S.O.

Hunter, D. 1975 *The diseases of occupations*. London: English Universities Press.

Jones, D. M., Bennett, D. & Elgar, K. E. 1978 Death of owls traced to insecticide-treated timber. *Nature, Lond.* **272**, 52.

Kates, R. W. 1978 *Risk assessment of environmental hazard*. (SCOPE Report No. 8.) (In the press.)

Kinnersley, P. 1974 *The hazards of work and how to fight them*. London: Pluto.

Kletz, T. A. 1976 The application of hazard analysis to risks to the public at large. In *Proceedings, World Congress of Chemical Engineering*. Amsterdam, July 1976, session A5.

Kletz, T. A. 1977 What risks should we run? *New Scient*. **74**, 320–322.

Knowles, J. H. 1977 The responsibility of the individual. *Daedalus, Boston, Mass*. **106**, 57–80.

Korte, F. 1977 Occurrence and fate of synthetic chemicals in the environment. In *Evaluation of toxicology data for protection of public health* (ed. W. J. Hunter & J. G. P. M. Smeets), pp. 235–246. Oxford: Pergamon.

Lassiter, D. V. 1977 In *Occupational carcinogenesis in environmental cancer* (ed. H. F. Kraybill & Mehlman M. A.), pp. 63–86. New York: Wiley.

Lederberg, J. 1974 A systems analytic viewpoint. In *How safe is safe* (ed. D. E. Koshland), pp. 66–94. Washington: National Academy of Sciences.

Lowrance, W. W. 1976 *Of acceptable risk*. Los Altos, California: W. Kaufmann.

Lundqvist, L. J. 1977 The politics of determining socially acceptable levels of risk in Sweden and the United States. In *Proceedings of Seminar on Environmental Risk Assessment*, Tihanyi, Hungary, June 1977. (SCOPE Project 7).

M.A.F.F. 1971 *Ministry of Agriculture, Fisheries and Food Pesticide Safety Precautions Scheme*. Pesticides Branch, Great Westminster House, London, S.W.1., U.K.

Mark, R. K. & Stuart, D. E. 1977 Disasters as a necessary part of benefit–cost analysis. *Science, N.Y.* **197**, 1160–1162.

McLean, A. E. M. 1972 In *Is it safe to live?*, vol. 4, pp. 331–335. London: Medicine (Poisoning).

McLean, A. E. M. 1974 Prevention of paracetamol poisoning. *Lancet* ii, 729.

McLean, A. E. M. 1978 Pesticides [reply to Gillespie *et al*. 1978]. *New Scient*. **78**, 40.

McLean, A. E. M. & Day, P. A. 1975 Effect of diet on the toxicity of paracetamol and the safety of paracetamol–methionine mixtures. *Biochem. Pharmacol*. **24**, 37–42.

McLean, A. E. M. & McLean, E. K. 1966 The effect of diet and DDT on microsomal hydroxylating enzymes and on the sensitivity of rats to carbon tetrachloride poisoning. *Biochem. J*. **100**, 564–570.

McLean, A. E. M. & Marshall, A. K. 1971 Reduced carcinogenic effects of aflatoxin in rats given phenobarbitone. *Br. J. exp. Path*. **52**, 322–329.

Mitchell, J. R., Potter, W. Z., Hinson, J. A., Snodgrass, W. R., Timbrell, J. A. & Gillette, J. R. 1975 Toxic drug reactions. In *Concepts in biochemical pharmacology*, vol. 3 (ed. J. R. Gillette & J. R. Mitchell), pp. 382–419. Berlin: Springer-Verlag.

Pochin, E. E. 1975 The acceptance of risk. *Br. med. Bull*. **31**, 184–189.

Registrar General 1971 *Decennial supplement England and Wales* 1961. *Occupational mortality tables*. London: H.M.S.O.

Registrar General 1974 *Report for England and Wales* 1972, Part 1 (Tables: medical table 13). London: H.M.S.O.

Registrar General 1978 *Occupational mortality* 1970–1972. London: H.M.S.O.

Royal Commission on Environmental Pollution 1976 *6th Report*. London: H.M.S.O.

Rycroft, R. J. G., Cronin, E. & Calman, C. D. 1976 Canadian sheet dermatitis. *Br. med. J*. ii, 1175.

Science Council of Canada 1977 *Policies and Poisons*. (Report no. 28.) Ottawa, Canda: Science Council of Canada.

Siekevitz, P. 1970 Scientific responsibility. *Nature, Lond*. **227**, 1301–1303.

Sinclair, C., Marstrand, P. & Newick, P. 1972 *Innovation and human risk*. London: Public Centre for Study of Industrial Innovation.

Slovic, P., Fischhoff, B. & Lichtenstein, S. 1976 Cognitive processes and societal risk taking. In *Cognition and social behaviour* (ed. J. S. Carroll & J. W. Payne), pp. 165–184. New York: Wiley.

Smith, H. V. & Spalding, J. M. K. 1959 Outbreak of paralysis in Morocco due to ortho cresyl phosphate poisoning. *Lancet* ii, 1019–1021.

Smith, R. J. 1977 Public gains access to pesticide safety data. *Science, N.Y.* **197**, 1346–1347.

Starr, C. 1969 Social benefit versus technological risk. *Science, N.Y.* **165**, 1232–1238.

Starr, C., Rudman, R. & Whipple, C. 1976 Philosophical basis for risk analysis. *A. Rev. Energy* **1**, 629–662.

Tiller, J. R., Schilling, R. S. F. & Morris, J. N. 1968 Occupational toxic factors in mortality from coronary heart disease. *Br. med. J.* iv, 407–411.

Van Rensburg, S. J., Van Der Watt, J. J., Purchase, I. F. H., Continho, L. P. & Markham, R. 1974 Primary liver cancer rate and aflatoxin intake in a high cancer area. *S. Afr. med. J.* **48**, 2508a–2508d.

Westlake, M. 1977 Refuelling the engine of economic growth.*The Times*, October 25, p. 19.